RURAL BUILDING

VOLUME ONE

RURAL BUILDING

VOLUME ONE

A Reference Book

INTERMEDIATE TECHNOLOGY PUBLICATIONS 1995

Published by Intermediate Technology Publications Ltd
103–105 Southampton Row, London WC1B 4HH, UK
Under licence from TOOL, Sarphatistraat 650, 1018 AV, Amsterdam, Holland

© TOOL

A CIP catalogue record for this book
is available from the British Library

ISBN 1 85339 310 X

Printed in Great Britain by SRP, Exeter, UK

PREFACE

This official text book is designed purposely to meet the needs of trainees who are pursuing rural building courses in various training centres administered by the National Vocational Training Institute.

The main aim of this book is to provide much needed trade information in simple language and with illustrations suited to the understanding of the average trainee.

It is the outcome of many years of experiment conducted by the Catholic F.I.C. brothers of the Netherlands, and the German Volunteer Service instructors, in simple building techniques required for a rural community.

The National Vocational Training Centre is very grateful to Brothers John v. Winden and Marcel de Keijzer of F.I.C. and Messrs. Fritz Hohnerlein and Wolfram Pforte for their devoted service in preparing the necessary materials for the book; we are also grateful to the German Volunteer Service and the German Foundation For International Development (DSE) - AUT, who sponsored the publication of this book.

We are confident that the book will be of immense value to the instructors and trainees in our training centres.

DIRECTOR: National Vocational Training
Institute, Accra

© Copyright
by Stichting Kongregatie F.I.C.
Brusselsestraat 38
6211 PG Maastricht
Nederland

Alle Rechte vorbehalten
All rights reserved.

INTRODUCTION TO A RURAL BUILDING COURSE

Vocational training in Rural Building started in the Nandom Practical Vocational Centre in 1970. Since then this training has developed into an official four year course with a programme emphasis on realistic vocational training.

At the end of 1972 the Rural Building Course was officially recognised by the National Vocational Training Institute. This institute guides and controls all the vocational training in Ghana, supervises the development of crafts, and sets the examinations that are taken at the end of the training periods.

The Rural Building programme combines carpentry and masonry, especially the techniques required for constructing housing and building sanitary and washing facilities, and storage facilities. The course is adapted to suit conditions in the rural areas and will be useful to those interested in rural development, and to farmers and agricultural workers.

While following this course, the instructor should try to foster in the trainee a sense of pride in his traditional way of building and design which is influenced by customs, climate and belief. The trainee should also be aware of the requirements of modern society, the links between the old and new techniques, between traditional and modern designs -- and how best to strike a happy medium between the two with regard to considerations like health protection, storage space, sewage and the water supply. The trainee should be encouraged to judge situations in the light of his own knowledge gained from the course, and to find his own solutions to problems; that is why this course does not provide fixed solutions but rather gives basic technical information. The instructor can adapt the course to the particular situation with which he and the trainee are faced.

This course is the result of many years of work and experimentation with different techniques. The text has been frequently revised to serve all those interested in Rural Development, and it is hoped that this course will be used in many vocational centres and communities. It is also the sincere wish of the founders of this course that the trainees should feel at the completion of their training that they are able to contribute personally to the development of the rural areas, which is of such vital importance to any other general development.

We are grateful to the Brothers F.I.C., the National Vocational Training Institute and the German Volunteer Service for their assistance and support during the preparation of this course.

                                        Bro. John v. Winden (F.I.C.)
                                        Wolfram Pforte (G.V.S.)
                                        Fritz Hohnerlein (G.V.S.)

LAY-OUT OF THE RURAL BUILDING COURSE

The Rural Building Course is a block-release-system course, which means that the trainee will be trained in turn at the vocational centre and at the building site. The period of training at the centre is called "off-the-job" training, and the period on the building site is called "on-the-job" training. Each will last for two years, so that the whole course will take four years and will end with the final test for the National Craftsmanship Certificate.

BLOCK RELEASE SYSTEM

YEAR	TERM 1	TERM 2	TERM 3
1	X	X	X
2	O	O	O
3	O	X	O
4	X	O	X

X = OFF-THE-JOB TRAINING
O = ON-THE-JOB TRAINING

The total "off-the-job" training period is approximately 76 weeks, each week 35 hours. During this training about 80% of the time is spent on practical training in the workshop. The remaining 20% of the time is devoted to theoretical instruction.

The total "on-the-job" training period is approximately 95 weeks, each week 40 hours. During this period the trainee does full-time practical work related to his course work. In addition some "homework" is assigned by the centre and checked by the instructors.

A set of books has been prepared as an aid to the theoretical training:

        A - Rural Building, Basic Knowledge (Form 1)
        B - Rural Building, Construction (Forms 2, 3, 4)
        C - Rural Building, Drawing Book (Forms 1, 2, 3, 4)
        D - Rural Building, Reference Book

All these books are related to each other and should be used together. The whole set covers the syllabus for Rural Building and will be used in the preparation for the Grade II, Grade I, and the National Craftsmanship Certificate in Rural Building.

CONTENTS

INTRODUCTION

GENERAL RULES FOR SAFETY

| PART 1: RURAL BUILDING TOOLS | 1 |

LAYING TOOLS — 3
- The trowel

STRAIGHTENING TOOLS — 5
- Spirit level / Plumb bob / Mason line / Straight edge

MEASURING AND MARKING TOOLS — 11
- Folding rule / Zig-zag rule / How to read rules / Pencils / The mason square / The large square / The sliding bevel / Block gauges

CUTTING TOOLS — 15
- The block scutch / The pointed chisel / The stone breaking tool / The club hammer

FINISHING TOOLS — 17
- The pointing trowel / The big wood float / The small wood float

SITE TOOLS — 19
- Cutlass / Hoe / Shovel / Pick-axe / Rammer / Hacksaw / Bolt cutter / Headpan

SAFETY GEAR — 21
- Safety goggles

SITE EQUIPMENT — 23
- Screen / Bucket / Rope / Wheelbarrow / Bending plate / Watering can / Water level / Boning rods / Strike board / "Cinva Ram" block press / "Tek" block press / Sandcrete block machine / Spatterdash / Wooden mould / Bullock cart / Concrete mixer

THE WORKPLACE — 33

THE WORK BENCH — 33
- Bench hooks / The holdfast / The bench stop

THE TRY SQUARE — 37
- How to use the try square / How to test the try square

THE MARKING GAUGE · 39
- How to set the gauge / How to use the gauge

THE CLAWHAMMER · 41
- How to pull nails / How to drive nails

THE MALLET · 41

HANDSAWS · 43
- The ripsaw / The crosscut saw / The backsaw / How to use
  a handsaw / How to rip boards

THE CHISEL · 47
- The firmer chisel / The mortice chisel / How to use the chisel /
  Chiselling with the grain / Chiselling across a board / Using the
  mallet / Chamfering / Morticing / Cutting curves with the chisel

PLANES · 53
- The jack plane / The smoothing plane / How to set the cap iron /
  How to set the cutting iron / Planing / Stuffing

THE SHARPENING STONE · 61
- How to use the stone

OTHER TOOLS FOR EXTRACTING AND PUNCHING NAILS · · · · · 63
- Pincers / The crowbar / The nail punch

CLAMPS · 65
- The G-clamp / The metal sash clamp / The wooden sash clamp

THE BRACE · 67
- How to use the brace

BITS AND DRILLS · 71
- Auger bits / Centre bits / Expansion bits / Twist drills /
  Depth stops / Countersinks / Awls

RASPS AND FILES · 75
- How to use rasps and files

THE SCREWDRIVER · 77
- How to use the screwdriver

TOOLS FOR MEASURING AND MITRING ANGLES · · · · · · · · · · · 79
- The mitre square / The try and mitre square / The sliding
  bevel / The mitre box / The mitre block

ADDITIONAL PLANES FOR THE RURAL BUILDER · · · · · · · · · · · 81
- The router plane / The rebate plane

SANDPAPER   83
- How to use sandpaper

OPTIONAL TOOLS   85
- Other types of trowels / Steel float / Other planes

PART 2: MAINTENANCE OF TOOLS   89

CUTTING ANGLES   91

SHARPENING PLANE IRONS AND CHISELS   93
- Angle and shape of the cutting edge / How to grind plane irons
  and chisels on the grinding wheel / Grinding on the coarse side
  of the sharpening stone / How to hone plane irons and chisels

SHARPENING BITS AND DRILLS   101
- Sharpening the auger bit / How to sharpen the centre bit / How
  to sharpen the twist drill / How to sharpen awls

SHAPING SCREWDRIVERS   103
- Shaping cold chisels / Shaping block scutches

MAINTAINING WOODEN PLANES   105
- How to reface the sole / About the bedding of the cap iron /
  Fitting the wedge

MAINTENANCE OF SAWS   107
- General maintenance / The action of the teeth

ANGLE OF PITCH, SHAPE AND NUMBER OF SAW TEETH   109
- The ripsaw / The crosscut saw / The backsaw

SETTING HANDSAWS   111
- How to set a saw / Setting the ripsaw / Setting crosscut and
  backsaws

FILING HANDSAWS   115
- Using the sawfile / How to file the crosscut saw / How to file
  the backsaw / How to file the ripsaw

TOPPING A HANDSAW   119
- How to top a saw

| PART 3: RURAL BUILDING MATERIALS | 121 |

ABOUT WOOD IN GENERAL — 123
- The structure and growth of the tree / Hard and soft wood / The structure of wood / The path from standing tree to sawn timber / Conversion terms for solid timber / How to order timber

THE PROPERTIES OF WOOD — 129
- The moisture content of wood / Shrinkage

NATURAL SEASONING — 131
- Warping

TIMBER PILING — 133
- Layout of the wood stack / Making the stacks

SPECIFICATIONS OF WOODS WIDELY USED IN NORTHERN GHANA — 136
- Odum / Wawa

DEFECTS IN TIMBER — 139
- Knots / Twisted grain / Checks / Wane or waney edge / Deadwood

DISEASES IN TIMBER — 141
- Fungal diseases / Prevention of fungal diseases / Remedies for fungal attacks

INSECT ATTACK — 143
- Beetles / Prevention of beetle attack / Remedy for beetle attack Termites / Prevention of termite attack / Remedies for termite attack

PRESERVATION AND PROTECTION OF TIMBER AND MASONRY — 144
- Timber preservatives

AGGREGATES — 147
- Sand / Gravel / Broken stones / Quality and properties of aggregates

BINDING MATERIALS — 152
- Lime / Portland cement / History of cement / Storing binding materials

MORTAR — 158
- Types of mortar / Selecting the right kind of sand / Batching Mixing the mortar

CONCRETE 166
- Cement paste / Properties of concrete

PART 4: RURAL BUILDING PRODUCTS 171

REINFORCEMENT STEEL 171
- Types of reinforcement steel / Reinforcement mats / Binding wire

LANDCRETE BLOCKS 174
- Required materials / Required equipment / Testing and choosing the soil / Making blocks / Planning the work

SANDCRETE BLOCKS 191
- Making the blocks / Planning the work

DECORATIVE BLOCKS 193

VENTILATING BLOCKS 195

PRECAST CONCRETE MEMBERS 197
- Planning the work

GLUE 199

PROTECTIVE FINISHES 200
- Oil paint / Synthetic paint / White wash / Cement paint / PVA emulsion paint / Varnish / Paints and varnishes: purchasing

SHEET MATERIALS 203
- Plywood / Block board / Hard board / Chip board / Decorative laminated plastic

WOOD FASTENINGS 207
- Nails / Ordering nails / Screws / Ordering screws / Bolts and nuts / Coach bolt / Washers / Spring washer / Anchors

DOOR AND WINDOW HARDWARE 217
- Louvre windows / Hinges / Locks and fittings

ROOF COVERINGS 230
- Corrugated aluminium sheets / Corrugated galvanized iron sheets / Corrugated asbestos cement sheets / How to order sheets / Ridge caps

APPENDIX I : TABLES OF FIGURES 232

APPENDIX II: GLOSSARY 241

BOOK INTRODUCTION

Rural Building Tools, Maintenance of Tools, Materials and Products is a reference book. This means that you should not read it through at once like a textbook, but use it when you need to look up information about certain tools, about the maintenance of a tool, or about a kind of building material or building product.

This book is divided into 4 parts:

PART 1: RURAL BUILDING TOOLS

This part of the book covers the basic tools needed in Rural Building and how to use them. It also treats a variety of site tools, site equipment and some optional tools.

PART 2: MAINTENANCE OF TOOLS

This section is about how to maintain the tools, so that they work better and last longer.

PART 3: RURAL BUILDING MATERIALS

This part deals with building materials; both the traditional ones and modern ones, that are used in Rural Building.

PART 4: RURAL BUILDING PRODUCTS

This part of the book covers the products such as reinforcement steel, blocks, paint, sheet materials, roofing sheets etc. used in Rural Building.

It is very important that you know all the technical terms, so if you come across a word or term that you don't understand you should look it up in the glossary at the end of this book, where most of the technical terms are explained. If you can't find the word in the glossary, write it down and ask your instructor to explain it.

The Tables of Figures in Appendix I are intended to help you to figure out the amounts of building materials that will be required for the planned building.

GENERAL RULES FOR SAFETY

1. Give all your attention to the job and don't distract others.
2. Be sensible in your behavior, don't play with tools or run about the building site or workshed.
3. Be alert, watch out for any dangerous situations, warn your colleagues, and report it to the person in charge.
4. If you are not sure of the correct way to use a tool, ask your instructor.
5. Make sure that your workpiece is safe and securely fastened in place before you start cutting or any work.
6. When cutting wood, guide the cutting tool in the correct way and keep your hands away from the cutting edge. Always cut away from yourself.
7. Wear safety goggles when cutting blocks, breaking concrete or grinding tools.
8. Carry tools with the pointed ends down.
9. When you finish working with a tool, clean it and return it to the toolbox.
10. Never throw or drop tools.
11. Keep the place tidy. A workplace scattered with tools is dangerous.
12. Maintain your tools, work only with clean and sharp tools.
13. A good quality, well maintained tool can do half the work for you ....

NOTES:

PART 1: RURAL BUILDING TOOLS

During thousands of years of development, people became aware that making certain things and doing certain jobs requires the use of special tools. Technical tasks could not be done with bare hands alone.

To make the things that they wanted, people were forced to design and make different tools for different jobs. For example, without tools like the plough, the farmer would not be able to feed his family. The plough was invented to make his work easier and to make sure that he could harvest a surplus of food for the benefit of the whole society. The plough was invented step by step and adapted to serve under different conditions.

In the same way, tools are very important in building. They enable people to shape all kinds of materials into useful articles and to make improved shelters to protect them from the weather and from enemies.

Early tools were the axe and the cutlass, which allowed men to cut wood for building instead of breaking it, and to make things like ladders, wheels and stools.

Now the Rural Builder uses more advanced tools and it is necessary for him to learn how to use and treat these tools well, because they are valuable. Even the most skilled craftsman can do little or nothing without the proper tools.

Since the Rural Building trade combines the crafts of carpentry and masonry, the Rural Builder's set of tools must also be a combination of masonry and carpentry tools.

This combined set of tools is limited and adapted for building in rural areas. It will enable the Rural Builder to construct perfectly well the kind of living quarters that are needed in the Northern and Upper regions of Ghana, starting from the foundation to the last nail of the roof construction.

Due to the structure of this course it is convenient to introduce the masonry tools first.

NOTES:

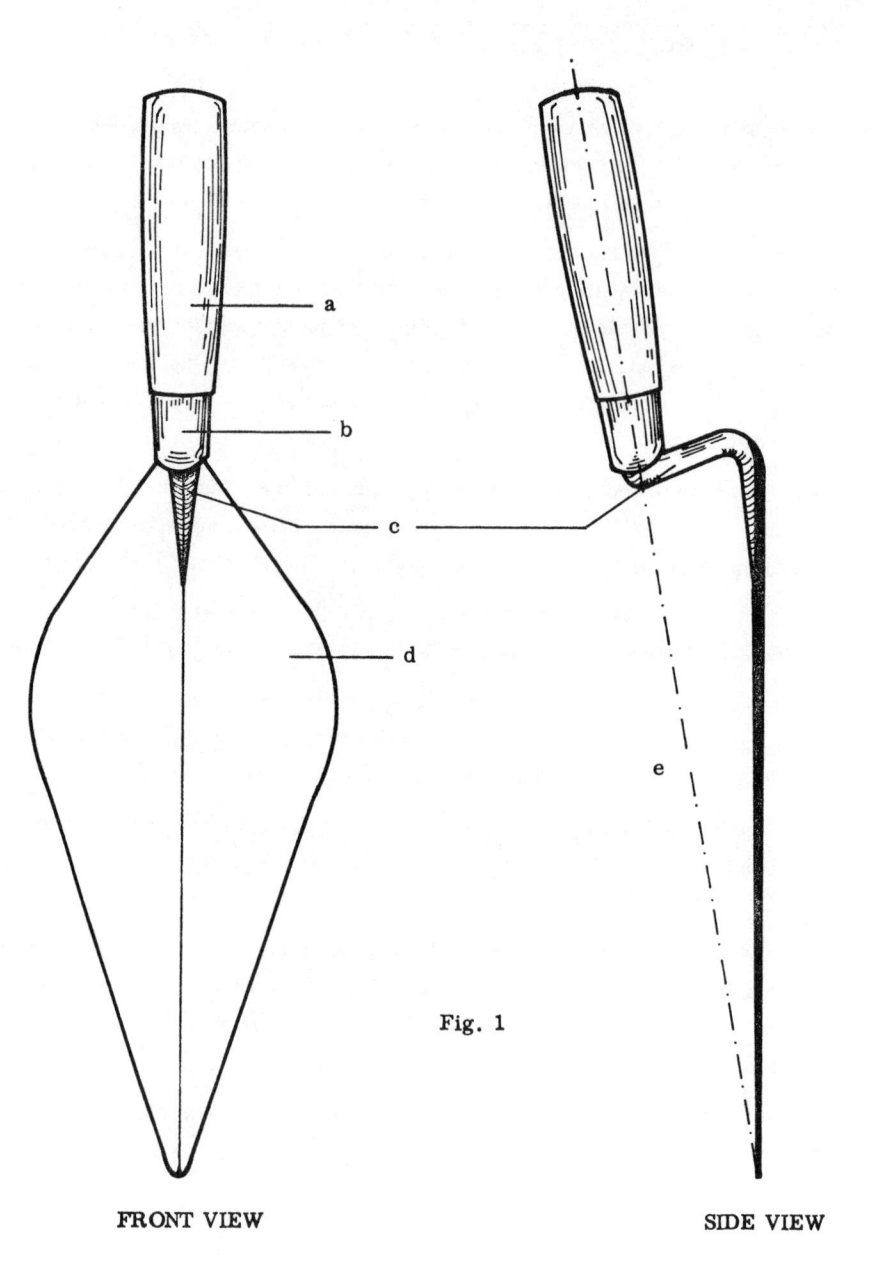

FRONT VIEW SIDE VIEW

Fig. 1

NEVER CUT SANDCRETE BLOCKS WITH A TROWEL, AS THIS WILL DENT THE EDGE OF THE BLADE.

N P V C		LAYING TOOLS.
2	TOOLS	

LAYING TOOLS

THE TROWEL

Of all the tools that a blocklayer uses, the brick trowel is by far the most important one, for it is almost continuously needed during the building construction.

Its main function is to pick up the mortar and to spread it to an even thickness in preparation for laying the blocks. Apart from its use in the trimming of landcrete blocks, the trowel is needed for any work where mortar or concrete is worked up.

NOTE: Never cut sandcrete blocks with a trowel, as this will dent the edge of the blade.

The long narrow-bladed trowel shown in Fig. 1 is very popular in this country and it is most frequently used for laying blocks and trowelling floorscreeds.

This trowel consists of a wooden handle (a) connected by a ferrule (b) to the shank (c) which joins the steel blade (d). The size of the blade ranges from 23 to 36 cm in length; this dimension being measured from the back of the shank to the tip of the blade, while the width varies from 9 to 13 cm.

The extended axle line of the handle (e) should line up with the tip of the blade in order to provide the best handling. This applies to all types of brick trowels.

When you buy a trowel, make sure that the blade is of a good quality steel. You can judge this quality by knocking your fingernail on the blade. The higher the sound the harder the steel.

The blade should also be able to bend slightly and return to its original position. If not, the blade is too soft.

NOTES:

WOODEN SPIRIT LEVEL   Fig. 1

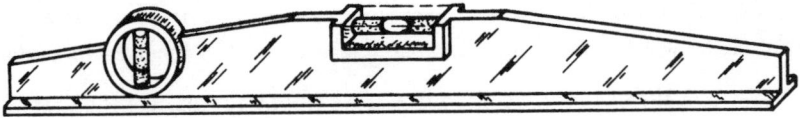

METAL SPIRIT LEVEL   Fig. 2

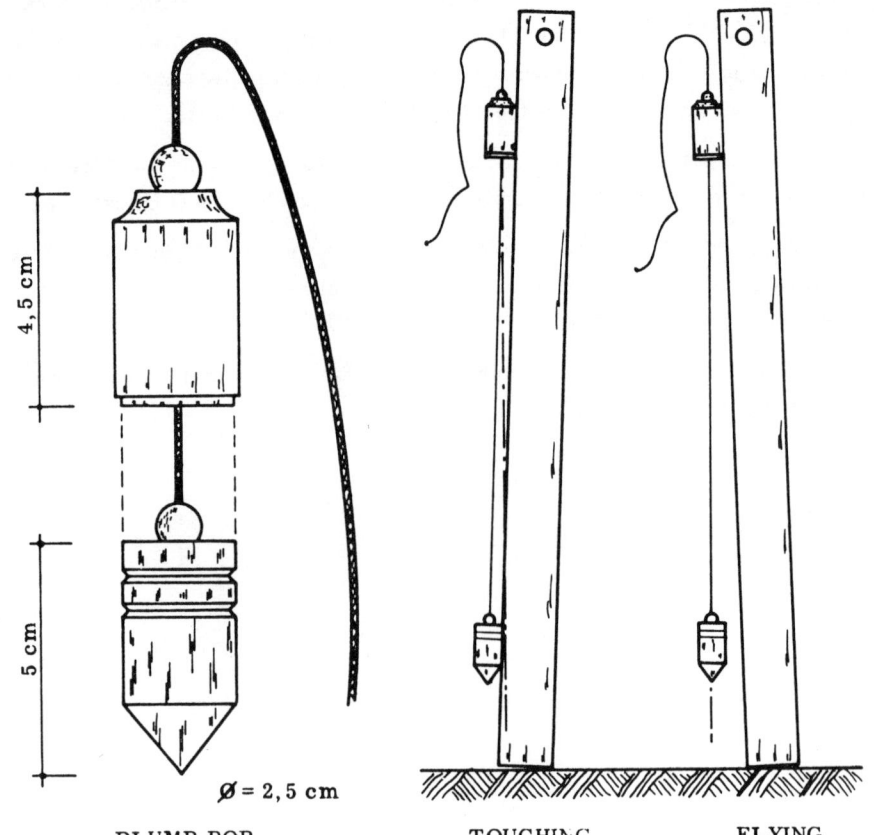

PLUMB BOB   TOUCHING   FLYING

Ø = 2,5 cm

N P V C		STRAIGHTENING TOOLS.
4	TOOLS	

STRAIGHTENING TOOLS

There are four main straightening tools.

SPIRIT LEVEL

These are wooden or metal straight edges specially fitted with plastic tubes containing spirit and a bubble of air.

These tubes are set into the straight edge so that when it is placed across two points which are level to each other, the air bubble will be exactly in the centre of the tube. This position is clearly marked with lines inside the tube (Figs. 1 & 2).

In a similar way, a tube is set in the straight edge to read with the level held vertically, which enables you to plumb members over short distances. If the level is used in conjunction with a straight edge you can plumb or level over a longer distance.

To level a longer horizontal distance you cannot use a straight edge with a level. Instead you have to use a water level which will be explained later. To level a vertical distance which is longer than your straight edge you can use your plumb bob.

PLUMB BOB

This tool consists of a solid brass or metal cylinder with a pointed end, which is attached to a suspending line so that its tip is always pointing vertically down. Its upper part is a small wooden block with a hole drilled in its centre so that the line with the cylinder on it can be pulled up or lowered down through the hole.

The diameter of the wooden block is slightly greater than the diameter of the cylinder, so that the cylinder can move freely up and down without touching the workpiece. The dimensions of the plumb bob are shown in Fig. 3.

The main use of the plumb bob is as a more accurate replacement for the vertical spirit level and also to transfer points down vertically in marking. Both methods will be described later.

NOTES:

Fig. 1

MASON LINE

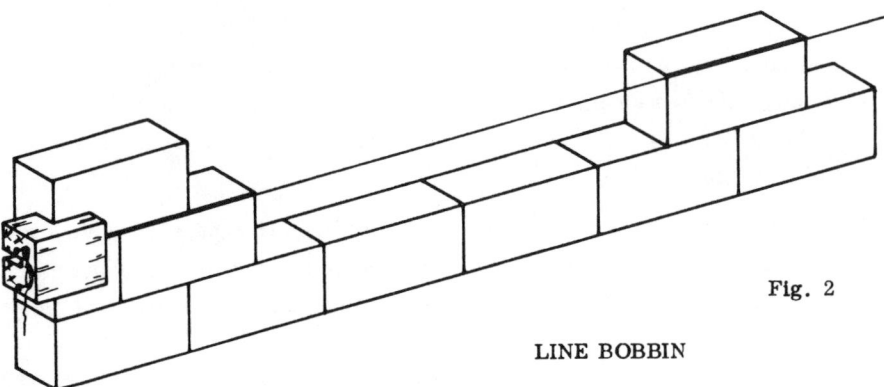

Fig. 2

LINE BOBBIN

There should be a trowel thickness ALLOWANCE between line and wall.

Piece of roofing sheet or stiff paper folded around line.

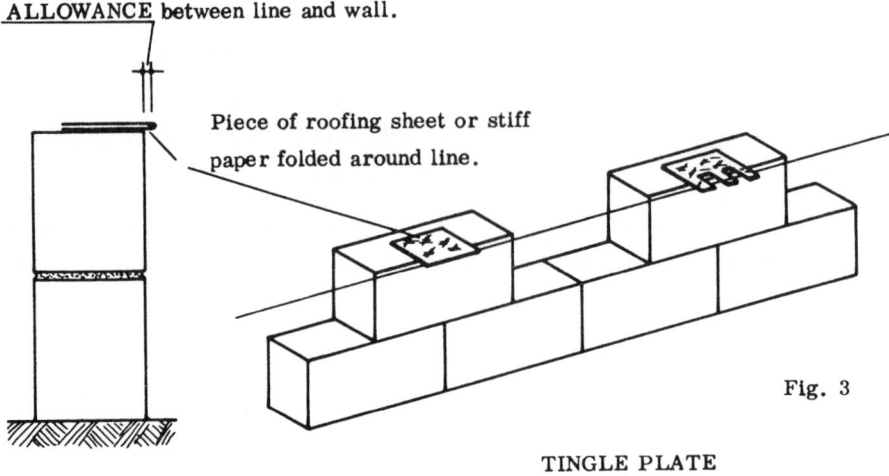

Fig. 3

TINGLE PLATE

N P V C		STRAIGHTENING TOOLS.
6	TOOLS	

MASON LINE

When building up walls between two quoins we employ the mason line, which is approximately 30 metres long, to ensure that the courses are straight and at the correct height (Fig. 1).

The line is tightened between two nails driven into the bed joints.

Mason lines are also used for setting out buildings, lining out frames for doors and windows and many other purposes where a straight line is needed for a guide over longer distances.

Instead of nails, so-called line bobbins may be used. These are hard wood blocks made to the size and shape indicated in Fig. 2. The line is stretched between opposite quoins, passed through the sawcut of each bobbin and wrapped around the projecting screws. Their uses will be explained later.

Line bobbins are preferred to nails, as they are easily adjusted to the required level and no holes need to be made in the bed joints.

In addition to the mason line, a tingle plate must be used if the distance between the quoins becomes too great and the line starts to sag. A tingle plate is made from thin metal and it is used to support the line in the middle to prevent sagging. The tingle plate must of course be set at the correct height (Fig. 3).

A tingle plate can easily be made from a piece of roofing sheet or any other sheet metal. Sometimes a piece of stiff paper is used for the purpose.

If the line breaks, it should be spliced and not tied with a knot, because a line full of knots will not be straight.

NOTES:

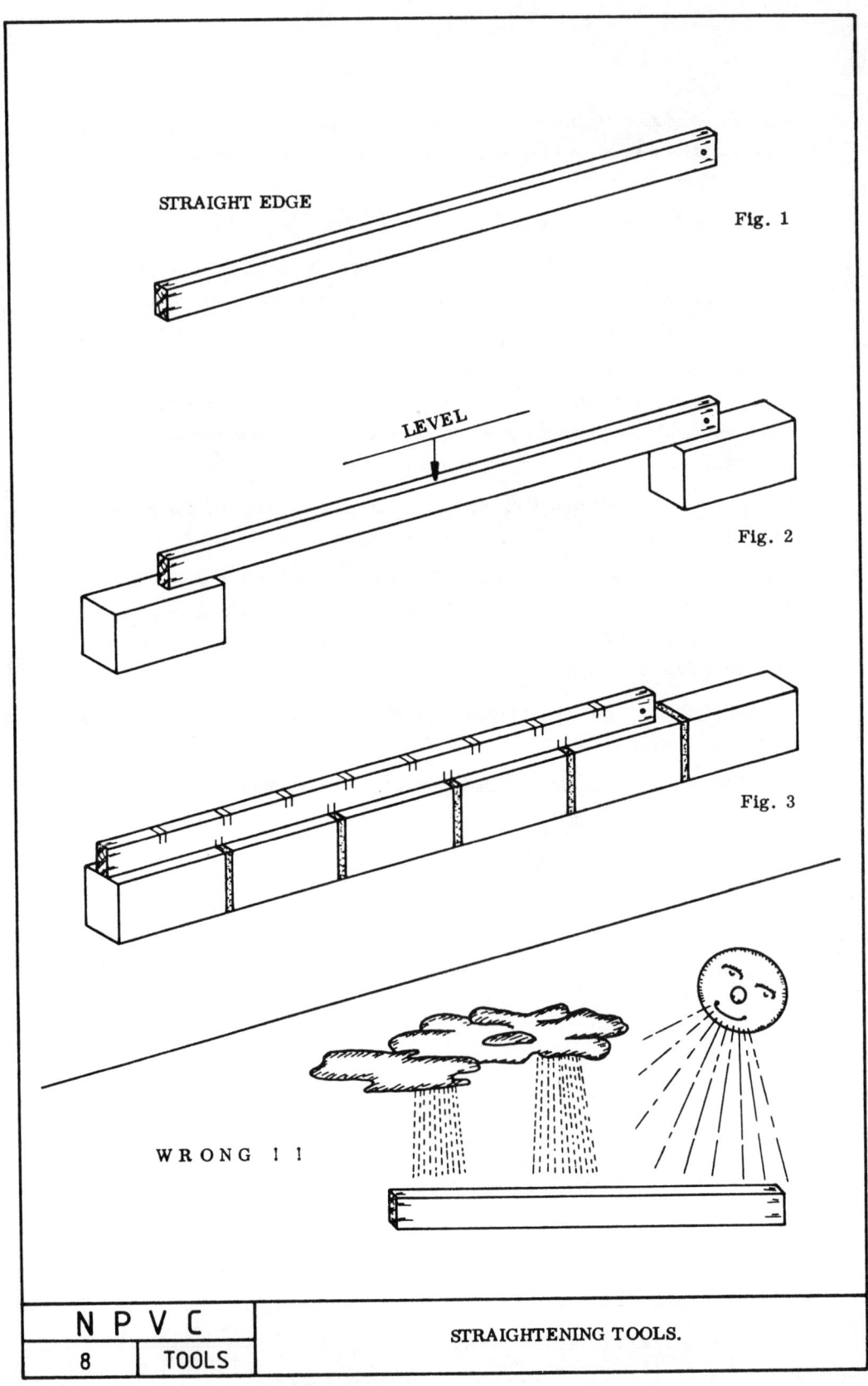

STRAIGHT EDGE

This is a planed piece of wood which should be well seasoned and dry to prevent it from bending and twisting. The dimensions of a straight edge are usually 2 to 2,50 m long, 7,5 cm wide and 2,5 cm thick; both edges must be perfectly straight and parallel (Fig. 1).

The straight edge is employed for testing masonry work either alone or in conjunction with the spirit level (Fig. 2).

Some straight edges are marked off with saw cuts to the required gauge; that is, one division is equal to the height of a block plus the joint; and, on the other edge, the length of a block plus the joint (Fig. 3).

Its wide range of further applications will be described as it is needed for certain constructions.

Do not allow a straight edge to dry out in the sun or to be soaked in water as this may cause it to bend or twist (Fig. 4). When you are finished using it, hang the straight edge in a protected place to keep it straight.

NOTES:

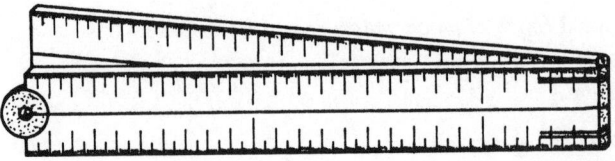

FOLDING RULE   Fig. 1

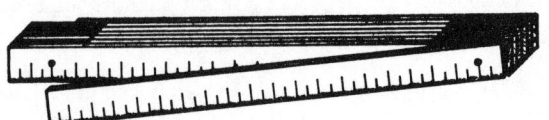

ZIG-ZAG RULE   Fig. 2

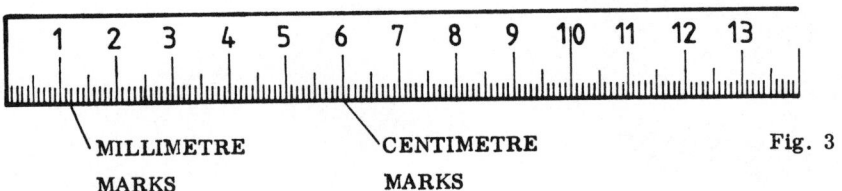

MILLIMETRE MARKS   CENTIMETRE MARKS   Fig. 3

PENCIL   Fig. 4

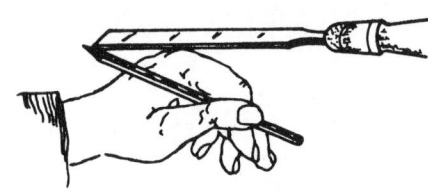

Fig. 5

N P V C	MEASURING AND MARKING TOOLS.
10 TOOLS	

MEASURING AND MARKING TOOLS

FOLDING RULE

The four-fold rule shown in Fig. 1 is made of four wooden, plastic or metal pieces which are held together by special hinges. It is one metre long and divided on both sides into millimetres and centimetres. It is used to find and check measurements as well as to mark out the work.

To make the rule operate more smoothly and last longer, put a drop of machine oil in the joints.

ZIG-ZAG RULE

A zig-zag rule is similar to a folding rule (Fig. 2). It is made out of the same materials but from pieces which are 20 cm long. As the hinges are different from those of the folding rule, be careful not to break it when opening and closing it. They come in lengths of 100 and 200 cm.

HOW TO READ RULES

The marks on a rule are of different lengths to make it easier to read accurately (look at the rule you use in class). The marks at each centimetre are the longest, the marks for 5 mm (1/2 cm) are medium long, while the millimetre marks are shortest (Fig. 3).

When measuring with the folding rule or the zig-zag rule, one must make sure that the rule is completely opened and straight. It is then held parallel to an edge, or at right angles to a face. If this is not done, the measurements you get will always be a little different from the correct ones.

PENCILS

For marking on wood, a hard lead pencil (H or 2H) is best (Fig. 4). The point should always be kept sharp, because using a blunt pencil can result in an inaccuracy of up to 2 mm.

Fig. 5 shows how to hold a pencil while sharpening it.

NOTES:

MEASURING AND MARKING TOOLS.

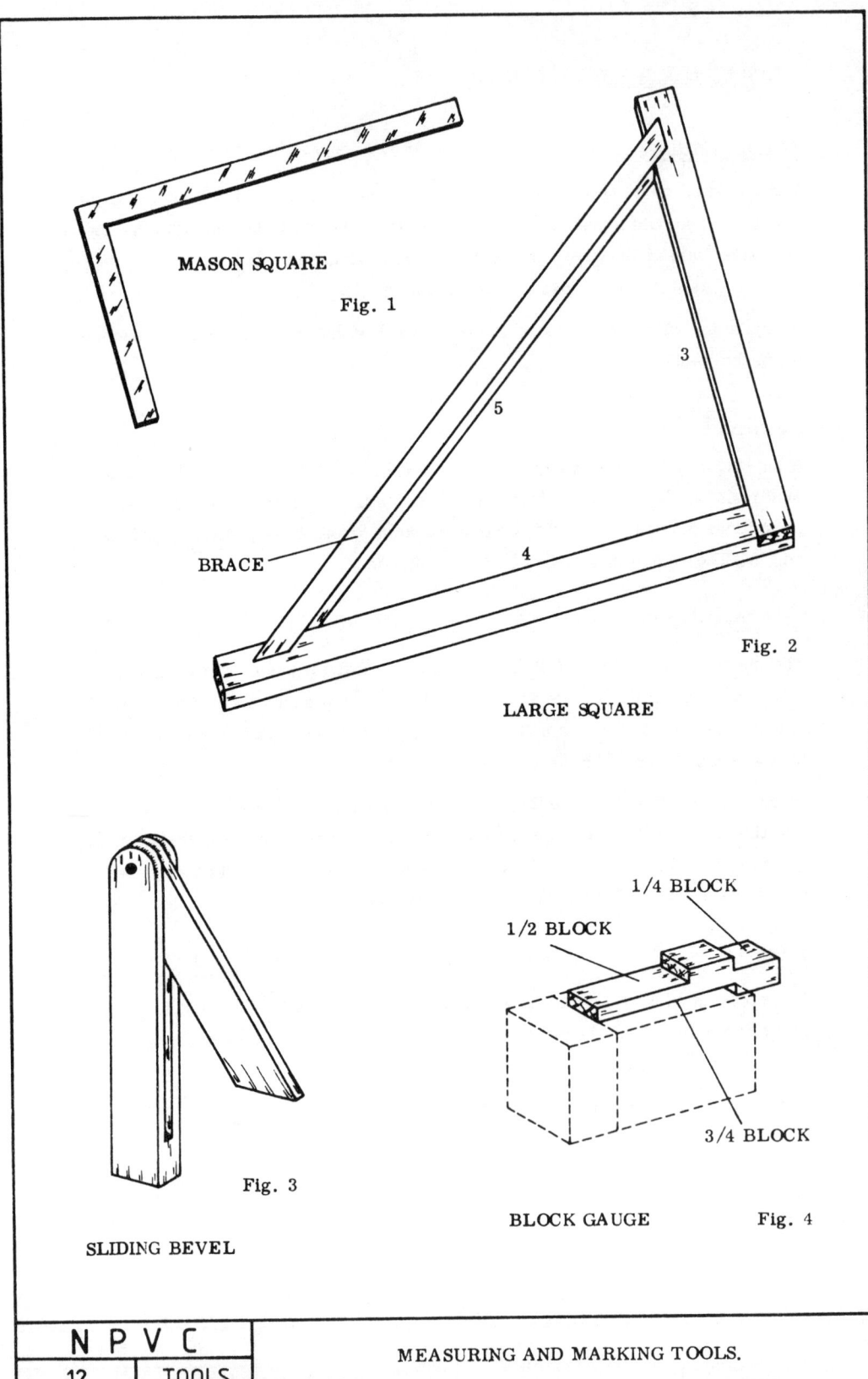

THE MASON SQUARE

The mason square is made from steel (Fig. 1). Measured along the outer edge, the short blade is 33 cm long and the long blade is 60 cm long. The blades are sometimes marked with millimetres, centimetres, and decimetres.

The mason square is used for setting out right angles as at quoins, and for testing corners during plastering.

When using the square, hold it either horizontally or vertically (not at an angle) to be sure of getting the correct angles.

THE LARGE SQUARE

This square is made entirely from wood (Fig. 2). To construct this large square which is made at the building site, use the 3-4-5 method and nail the boards together securely. A brace over the two legs ensures that the square remains at the correct angle. The square is used to test larger right angles.

THE SLIDING BEVEL

A sliding bevel (Fig. 3) can be made out of wood by the Rural Builder. The two legs are adjustable and held together by a small bolt with a wingnut to make it easy to adjust the bevel. It is used when you have to mark many blocks at a certain angle (also see page 78).

BLOCK GAUGES

These are pieces of wood cut to the size and shape indicated in Fig. 4. Block gauges may be used to mark off the sizes of 1/4, 1/2, or 3/4 blocks. Since the dimensions of landcrete blocks are different from sandcrete blocks, the trainee will make two different block gauges.

The gauges help the Rural Builder to work more efficiently when he is measuring blocks for cutting.

- NOTE: More tools for measuring and marking are included in the following sections with the carpentry tools.

NOTES:

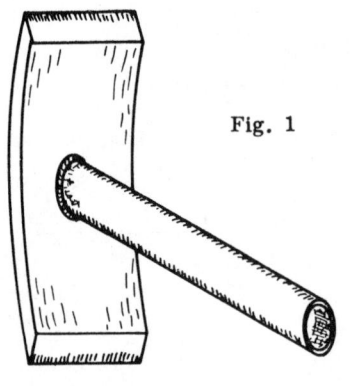

Fig. 1

BLOCK SCUTCH OR SCUTCHER

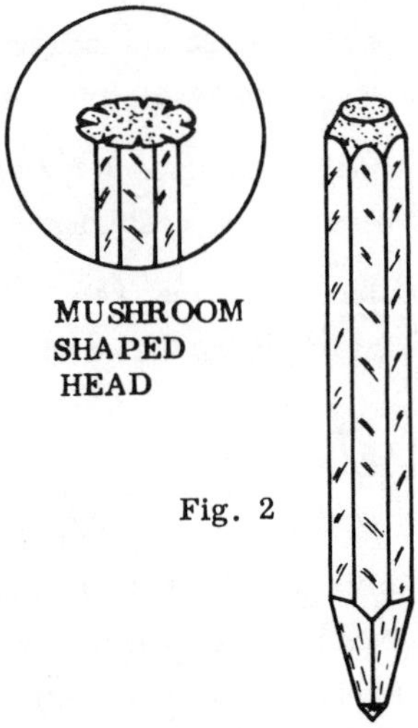

MUSHROOM SHAPED HEAD

Fig. 2

POINTED CHISEL

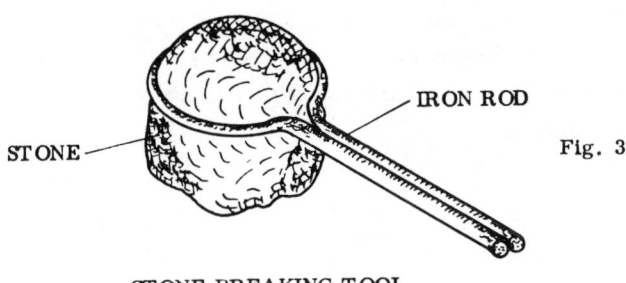

STONE — IRON ROD   Fig. 3

STONE BREAKING TOOL

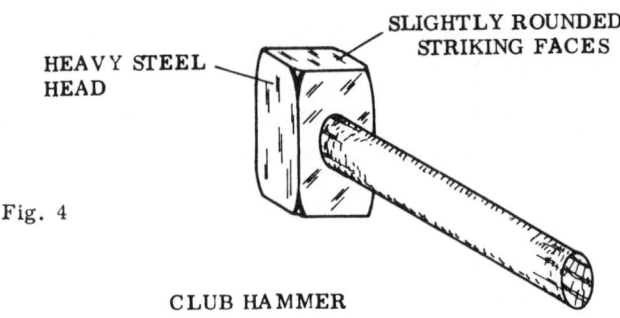

SLIGHTLY ROUNDED STRIKING FACES

HEAVY STEEL HEAD

Fig. 4

CLUB HAMMER

N P V C		CUTTING TOOLS.
14	TOOLS	

CUTTING TOOLS

THE BLOCK SCUTCH

This tool consists of a hard steel blade with two cutting edges, welded to the handle which is made of iron pipe (Fig. 1). It is used for cutting all sorts of blocks and dressing cut surfaces. The angle between the blade and handle should be 75 to 80 degrees, which increases the effectiveness of the blow. The handle is oval shaped to provide a better grip.

THE POINTED CHISEL (COLD CHISEL)

This is a forged steel rod with a hardened cutting tip and striking end. It is octagonally shaped to provide a better grip for the hand (Fig. 2). Cold chisels are available in different sizes and are used together with a club hammer.

The head of the cold chisel should never be allowed to become mushroom shaped, as this may result in badly cut hands or in a piece of steel breaking off and piercing someone's eye. Always wear your safety goggles when you use the chisel.

THE STONE BREAKING TOOL

A device like the one in Fig. 3 is a useful tool on the building site. It can be made locally and is used to break stones into smaller pieces needed for concrete work. Place the device on top of the piece of stone and press it down during the hammer blow.

THE CLUB HAMMER

This hammer has a heavy steel head with slightly rounded striking faces and it can weigh from 1 to 2 kg.

The head is fixed on a wooden handle which is 15 to 20 cm in length. The hammer is used to strike cold chisels and to break stones into smaller sizes. When using the hammer, make sure that the wedge that holds the handle in the head is firmly in position (Fig. 4).

NOTES:

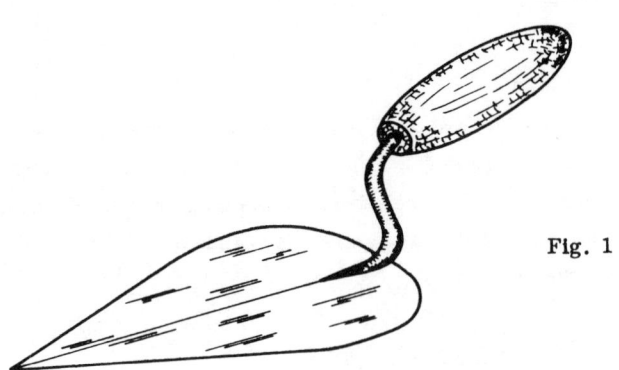

POINTING TROWEL Fig. 1

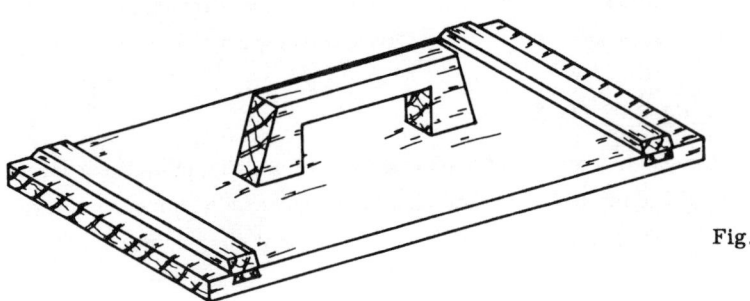

BIG WOOD FLOAT Fig. 2

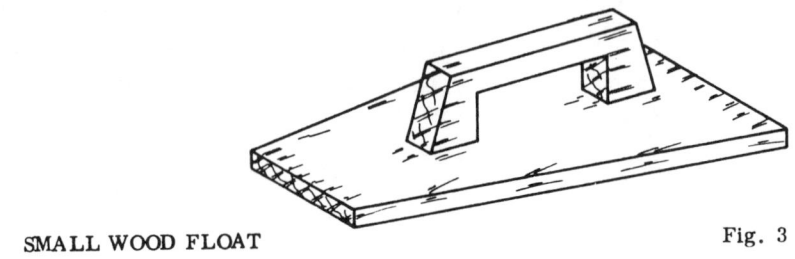

SMALL WOOD FLOAT Fig. 3

NPVC		FINISHING TOOLS.
16	TOOLS	

FINISHING TOOLS

THE POINTING TROWEL

Fig. 1 shows a trowel of almost the same shape as the brick trowel mentioned earlier, but smaller in its dimensions.

This pointing trowel is chiefly used for precision work such as finishing in general and the dressing of corners and edges in particular.

THE BIG WOOD FLOAT

This tool has a blade made of a soft wood like Wawa. It measures approximately 40 cm long and 25 cm wide. A handle made from hard wood is fixed to it with screws so that the blade can be replaced when necessary (Fig. 2).

Its main uses are to distribute an even thickness of mortar during plastering and to flatten concrete surfaces during floor construction.

THE SMALL WOOD FLOAT

The small wood float is constructed in the same way as the big one but with smaller dimensions, being approximately 25 cm in length and 15 cm in width.

As it is used mainly to give the plaster and floor surfaces a smoother finish, its blade may be made from hard wood (Fig. 3).

Because the wood float is made from wood it absorbs water from the wet mortar or cement during use, and it tends to warp. To prevent it from warping, keep the float under water when it is not in use so that all the sides are wet and the wood swells evenly.

NOTES:

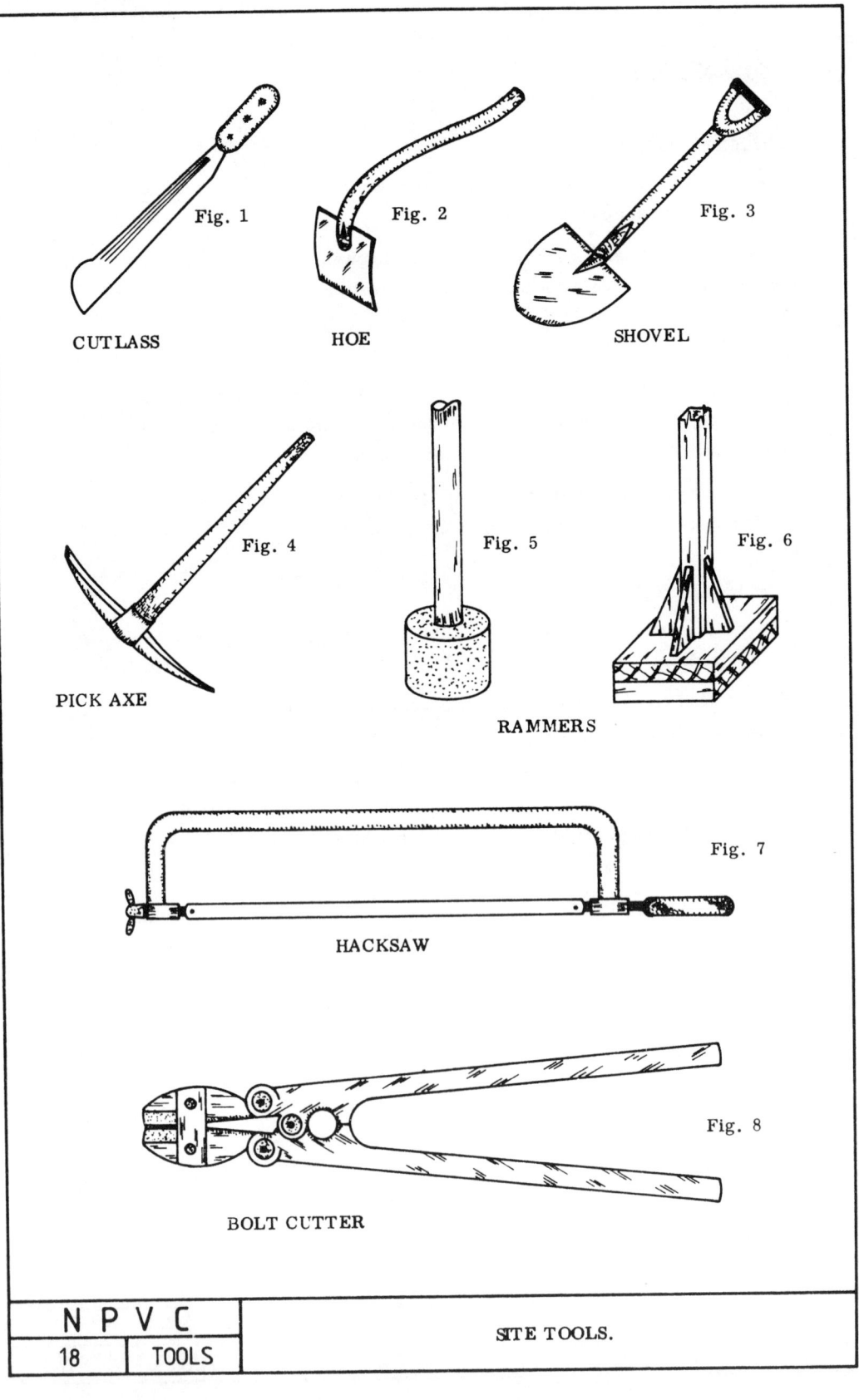

SITE TOOLS

CUTLASS

The cutlass (Fig. 1) is used for clearing the site and other general cutting work.

HOE

This farming tool is often used in Rural Building to excavate top soil (Fig. 2).

SHOVEL

There are various types of shovel-like tools. The most common type is the one with a round-nosed steel blade of about 25 by 30 cm, connected to a short wooden shaft that has a "D" or "Y" shaped handle at the end (Fig. 3).

Whether the shovel has a short or a long handle is a matter of personal preference or local custom. It has been observed that the short-handled one is more suitable for filling purposes and for moving light soil, while the long-handled shovel with a square steel blade is better for loading sand and for mixing.

PICK-AXE

This digging tool consists of a heavy steel head with one pointed end and one end with a chisel edge. The head is connected to a wooden shaft (Fig. 4). The pick-axe is used during excavation to break up hard rocky soils or loosen laterite etc.

RAMMER

Rammers are either made entirely out of wood or they have a wooden handle attached to a metal or concrete head (Figs. 5 & 6). They are used to compact soil or concrete.

HACKSAW

A hacksaw is a handsaw used for cutting metal. It consists of a steel blade tightly stretched in a metal frame. The blade is removable and other blades can be fixed in the frame for cutting asbestos-cement or other materials (Fig. 7).

BOLT CUTTER

The bolt cutter is a tool which is used to cut steel reinforcing rods up to 19 mm in diameter (Fig. 8).

Fig. 1

METAL HEADPAN

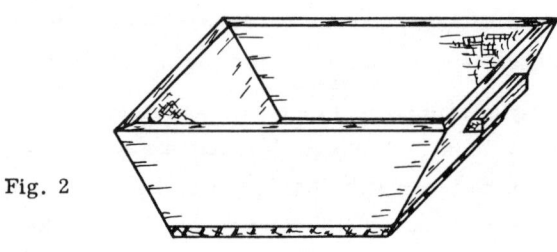

Fig. 2

WOODEN HEADPAN

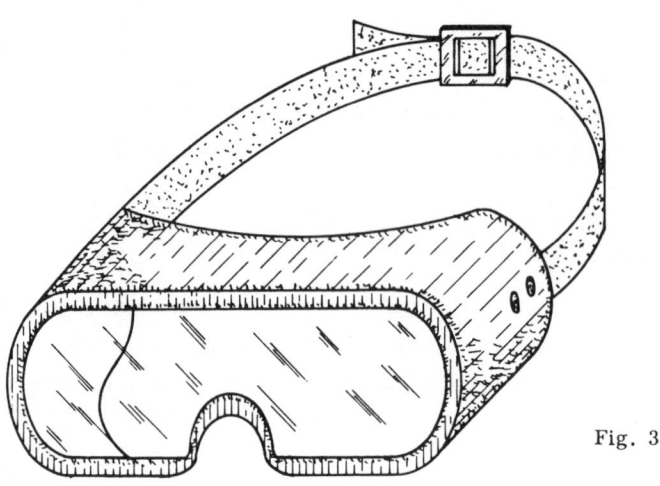

Fig. 3

SAFETY GOGGLES

N P V C	SITE TOOLS & SAFETY GEAR.
20	TOOLS

HEADPAN

Smaller quantities of mortar and concrete are kept and transported in headpans. These are round containers shaped like bowls and made from mild steel or sheet metal (Fig. 1).

If made locally from wood, the headpan will be square with slanting sides (Fig. 2).

The common headpan has a holding capacity of about 15 litres for liquids, or half a bag of cement (slightly heaped up). These figures indicate that the headpan can also be used as a measure.

If you make your own headpan from wood or metal, be sure that it has the same capacity as other headpans so that you can measure accurately with it.

SAFETY GEAR

SAFETY GOGGLES

These are made from plastic and are designed to protect the eyes during all kinds of cutting or grinding operations and where there is a lot of dust in the air (Fig. 3).

NOTES:

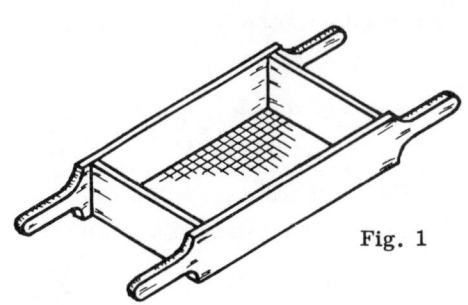

Fig. 1

SCREEN

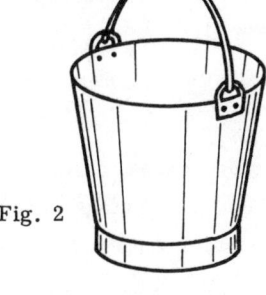

Fig. 2

BUCKET

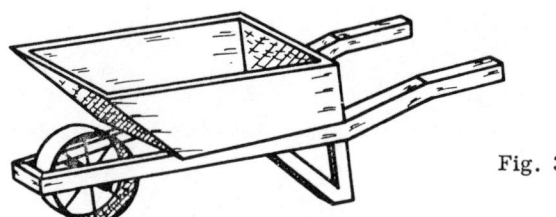

Fig. 3

WHEELBARROW

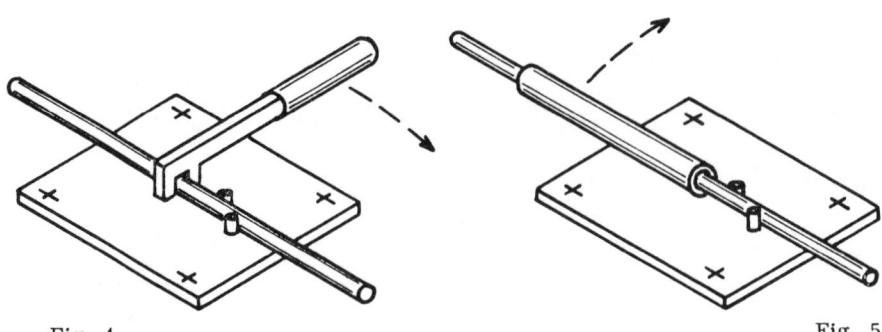

Fig. 4   Fig. 5

BENDING PLATE

N P V C		SITE EQUIPMENT.
22	TOOLS	

SITE EQUIPMENT

SCREEN

A screen is a rectangular frame with a wire mesh built into it for separating impurities or stones from sand (Fig. 1). A well-equipped building site will have two different screens: a larger mesh to separate out stones of a convenient size; and a smaller one to sieve sand that will be used for plastering.

BUCKET

Buckets are open containers that can be made from different materials like rubber, plastic, or galvanized iron (Fig. 2). The average bucket (size no. 28) has a volume of 10 litres and is used mainly for carrying water.

ROPE

Ropes used on the building site are usually made from hemp or nylon. Nylon ropes have a tendency to stretch when they are under strain, and this must be taken into consideration when you use this kind of rope during construction work.

WHEELBARROW

The wheelbarrow is a steel or wooden container with a single steel or rubber-tyred wheel in front. It is lifted and pushed forward by means of two hand-holds attached to the frame (Fig. 3).

BENDING PLATE

Our Rural Building equipment for bending iron rods simply consists of a baseplate with two steel pegs which are spaced according to the diameter of the rod to be bent, and a bending bar (Fig. 4). The bending bar is used to do the actual bending. This is a key-shaped tool with a slot in one side into which the rod fits. Each different diameter of rod needs its own bending bar. If a suitable bending bar is not available, a pipe can be used to do the job (Fig. 5).

WATERING CAN

The watering can (not illustrated here) is a container with a pouring spout, used for watering plants. On the building site it is often used to wet down newly poured concrete or freshly made sandcrete blocks.

Fig. 1

Fig. 2

WATER LEVEL

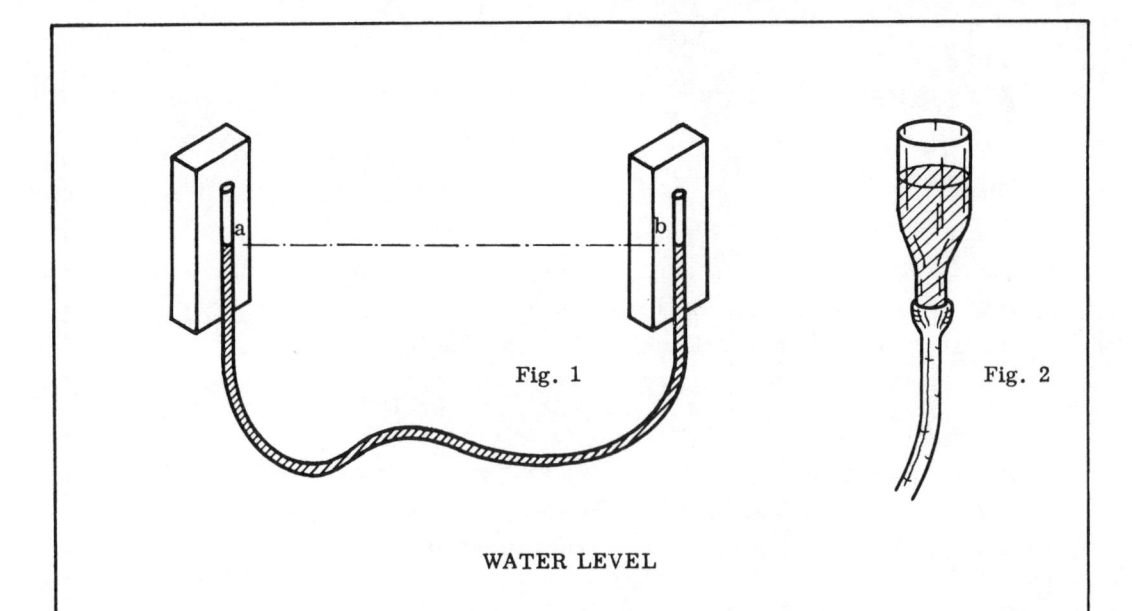

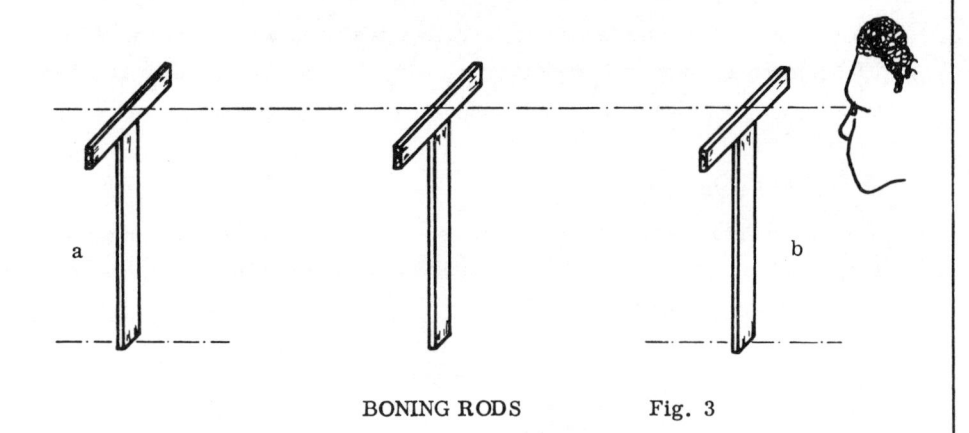

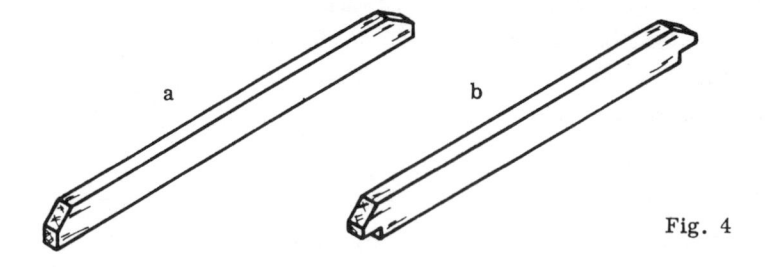

BONING RODS   Fig. 3

STRIKE BOARD

Fig. 4

N P V C		
24	TOOLS	SITE EQUIPMENT.

WATER LEVEL

This instrument is used for setting out levels on the site as well as to transfer and control levels over large distances. It consists of a transparent plastic tube filled with water (Fig. 1). The level of the water at one end of the tube (a) will be at exactly the same height as the level at the other end (b), provided that there is no air bubble in the tube and it is not buckled.

The water level enables us to level over large distances with a high degree of accuracy.

If there is no transparent plastic tube available and some rubber hose can be found, the Rural Builder can take two glass bottles, knock out the bottoms and fit the bottle necks to each end of the hose. This apparatus is then filled with water until the water is seen in the bottles. Levels can be read as easily with this device as with any other water level (Fig. 2).

BONING RODS

Boning rods are T-shaped wooden tools, usually 120 cm high and 20 cm wide at the top. They are used in sets of three to help the Rural Builder to level between two given points (Fig. 3).

Points a and b are marked with the water level and any point in between them can be obtained by using the third boning rod and sighting along the rods (Fig. 3).

STRIKE BOARD

Strike boards are made from well seasoned wood. They are similar to straight edges except that they are usually longer (Fig. 4).

Strike boards (a) are used to level off the screed on floors, or in the case of notched strike boards (b) to level off concrete before the screed is layed.

NOTES:

Fig. 1

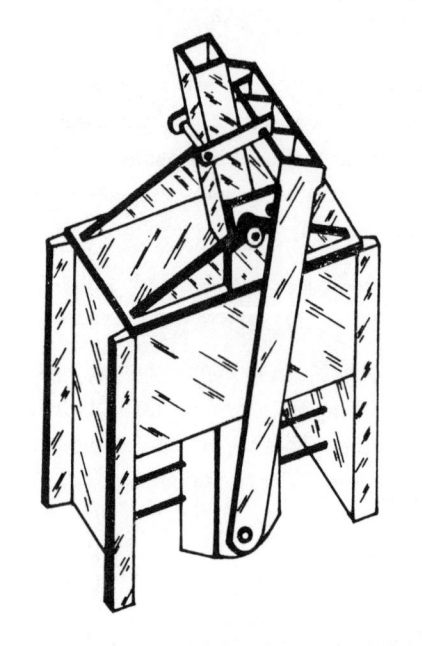

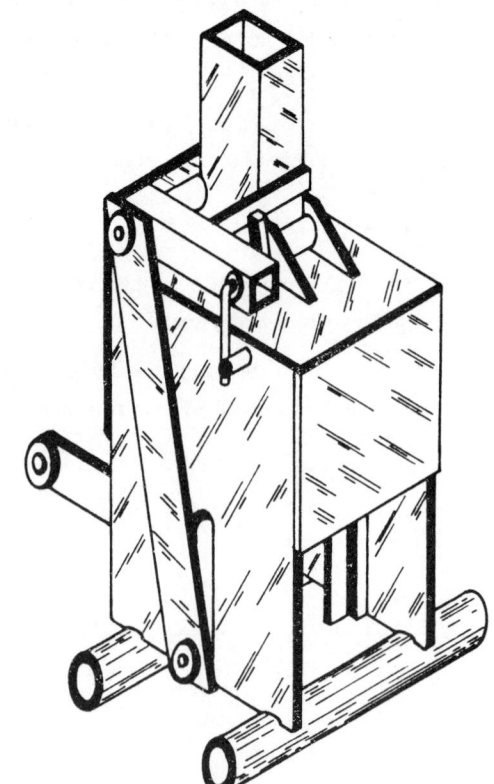

"CINVA RAM" BLOCK PRESS

"TEK BLOCK" PRESS

Fig. 2

N P V C		SITE EQUIPMENT.
26	TOOLS	

"CINVA RAM" BLOCK PRESS

The Cinva Ram block press is a simple low-cost machine which produces building blocks from common laterite. It consists of a mould in which a slightly moist soil/cement mixture is compressed (packed down) by a hand operated piston and lever system (Fig. 1).

Unlike sand-cement blocks made with a similar press, these blocks can be removed immediately from the press and stacked for curing without the use of a pallet underneath.

The machine is made from steel and it is tough, durable and will stand up to long and hard use. Little maintenance is needed except for oiling.

- SPECIFICATIONS: Size of blocks -- 10 x 15 x 30 cm
  Building unit -- 12 x 17 x 32 cm (2 cm joints).

"TEK" BLOCK PRESS

The Tek block press is similar to the Cinva Ram, except that this press produces blocks of larger dimensions (Fig. 2).

The Tek block press was designed at U.S.T. Kumasi and has been used successfully in the field for many years. The Tek block press can make blocks in any area where good laterite is available.

A moist soil/cement mixture is put into the steel mould box, pressed into a block and ejected (pushed out).

With the wooden handle, one man can put a lot of pressure on the block, so the blocks are very hard and long-lasting.

- SPECIFICATIONS: Size of blocks -- 14 x 22 x 29 cm
  Building unit -- 16 x 24 x 31 cm (2 cm joints).

- MAINTENANCE: Oiling the machine will make it work better and last longer. Oil or grease the moving parts at least twice a day when the machine is in use.

  Wipe the inside of the mould box with oil about every 10 blocks. This will make it easier to remove the blocks.

  If more than one man presses on the handle of the machine, the handle should break. If the handle is too big and strong, the machine will break instead of the handle. A broken machine is far more difficult to repair than a broken handle, therefore do not use any handle which is longer than 2,5 m.

  When the machine is not in use, paint it with oil to prevent rust.

SITE EQUIPMENT.

HAND OPERATED
SANDCRETE BLOCK MACHINE

Fig. 1

Fig. 2

PALLET

Fig. 3

SPATTERDASH APPARATUS

Fig. 4

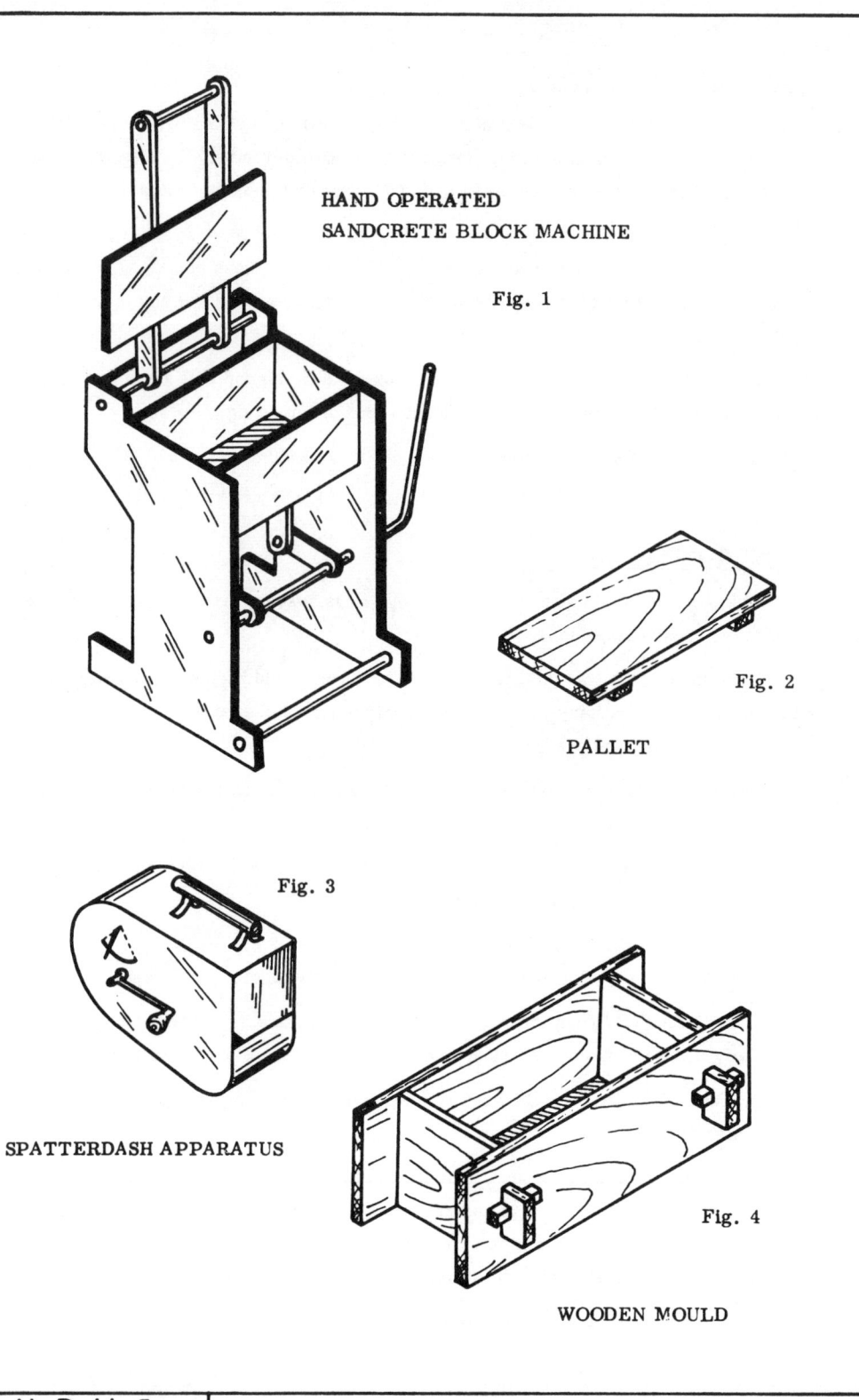

WOODEN MOULD

NPVC		
28	TOOLS	SITE EQUIPMENT.

HAND OPERATED SANDCRETE BLOCK MACHINE

This type of block machine consists mainly of a mould box with a movable bottom plate mounted into a supporting frame. The bottom plate is connected to a handle so that it can be raised (Fig. 1).

The lid is made of heavy material so that it presses down to compact the sandcrete. Sometimes additional weights are attached to the top of the lid to make it press harder.

After it is compacted, the block is pushed out by means of the handle at the side of the machine.

Unlike landcrete blocks, sandcrete blocks must be made on pallets, as they are too soft to be carried when they are freshly made (Fig. 2).

The inside measurements of the mould box are approximately: 46 cm in the length, 23 cm in the width, and 28 cm deep. Blocks of various shapes can be made with this machine by changing the height of the pallet or by using inserts.

SPATTERDASH APPARATUS

The spatterdash apparatus is used to give plaster or concrete an attractive appearance without the use of paint.

A slurry of sand and cement is placed in the apparatus (Fig. 3); when the handle is turned, the mixture is thrown against the wall, giving the surface a textured effect. Another handle adjusts the texture of the spatterdash from rough to fine.

WOODEN MOULD

A specially made wooden mould can be used when unusual blocks such as archblocks are needed or when there is no block-making machine available.

To make work easier and more accurate, the wooden mould should be made so that the sides of the mould can be removed easily and fixed back together by using wedges through tenons (Fig. 4).

NOTES:

SITE EQUIPMENT.

BULLOCK CART

Fig. 1

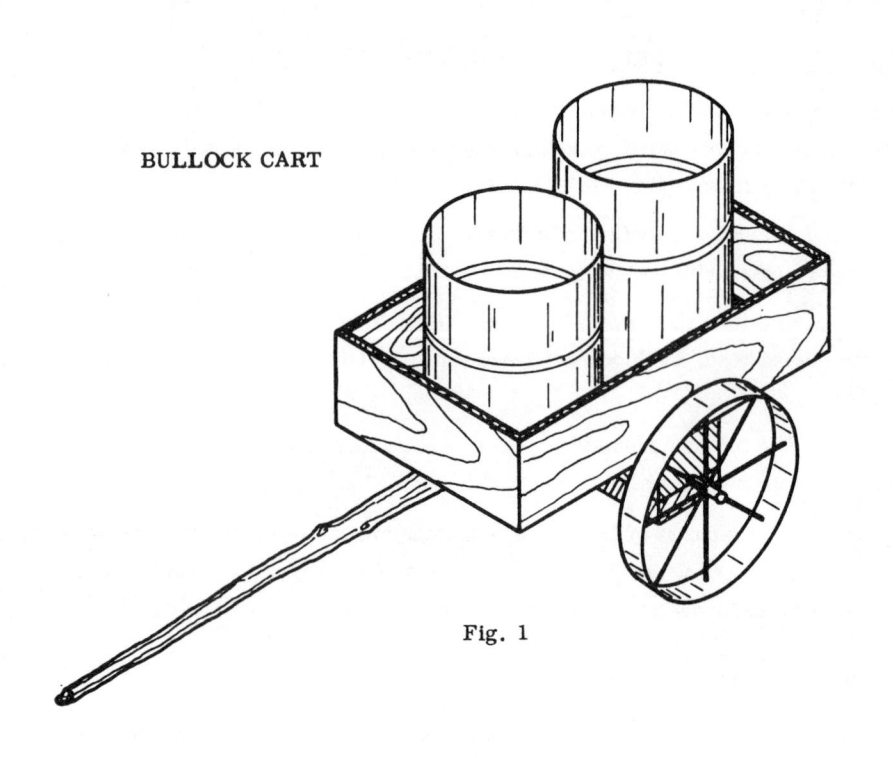

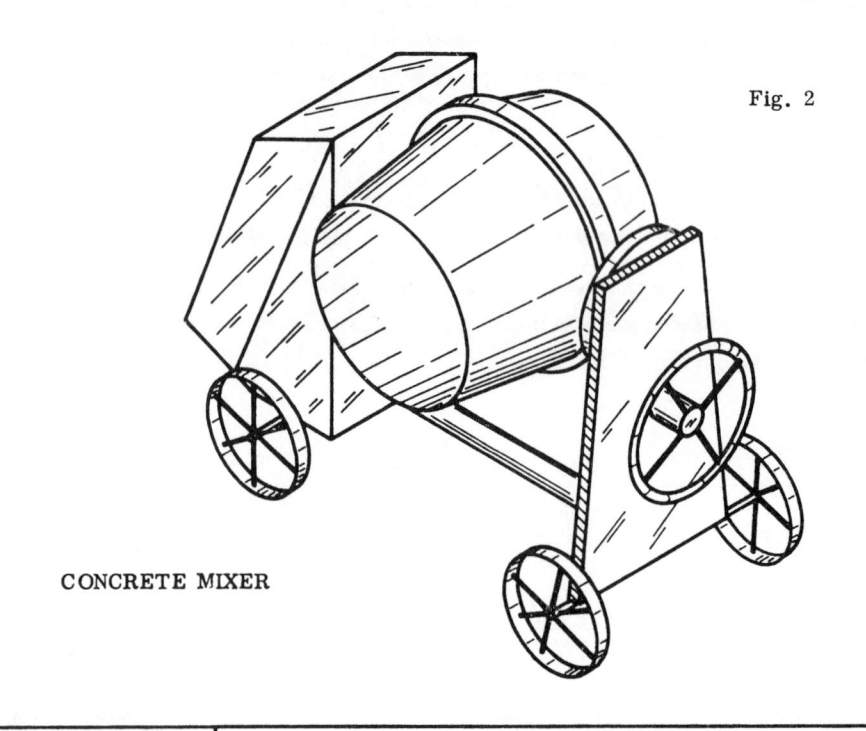

Fig. 2

CONCRETE MIXER

N P V C		
30	TOOLS	SITE EQUIPMENT.

BULLOCK CART

A bullock cart can transport water and limited quantities of building materials and it is therefore particularly useful at a Rural Building site. Bullock carts are made locally from wood and have steel rimmed wheels. The steel axles turn in wooden bearings which are soaked in oil. These bearings should be inspected and oiled regularly (Fig. 1).

CONCRETE MIXER

If there are few labourers and it is necessary to mix large amounts of concrete, a concrete mixer can be hired to do some of the work (Fig. 2).

The Rural Builder should know how to load the machine in the right sequence and make sure that the concrete is well mixed and has the correct content of water.

After the job is done the machine should be cleaned thoroughly with water and all sand and cement should be washed off from the drum and the frame of the machine.

A maintenance card will tell you how to maintain the machine and how often this needs to be done.

The engine needs special care and the Rural Builder should learn as much as he can about caring for it.

NOTES:

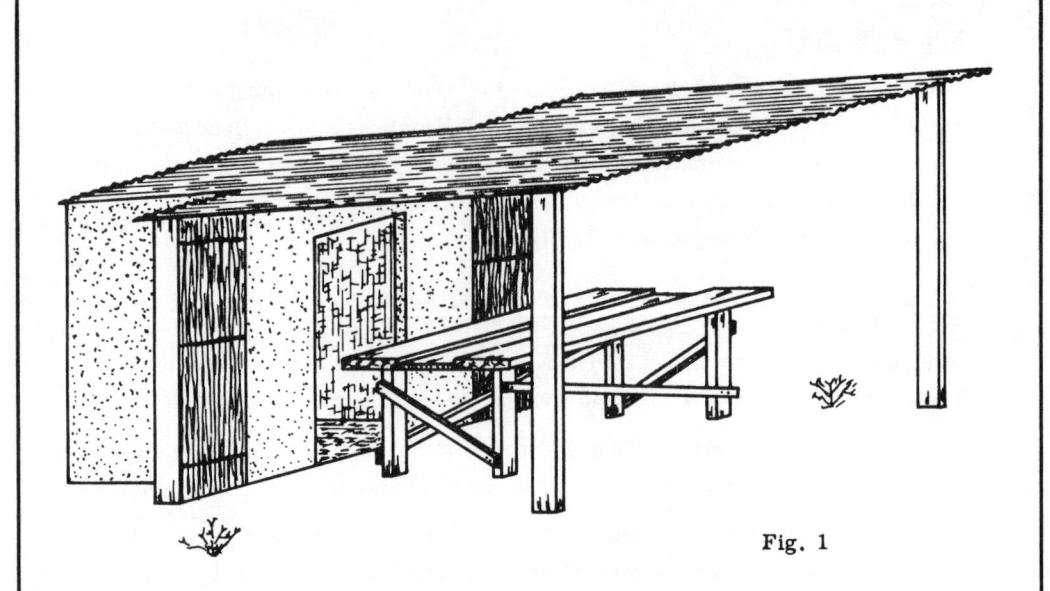

Fig. 1

THE WORKPLACE

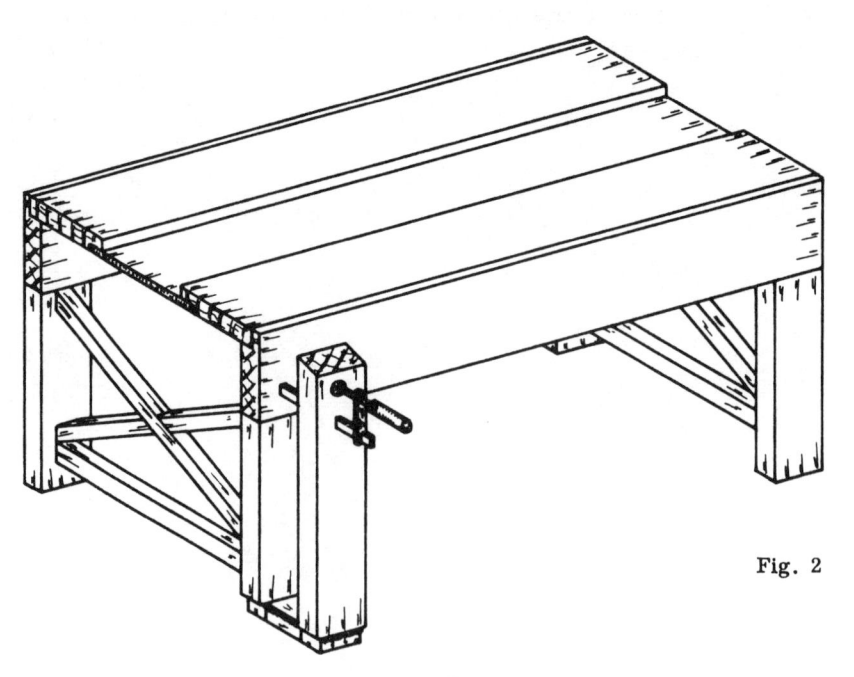

Fig. 2

THE WORK BENCH

N P V C		
32	TOOLS	THE WORKPLACE & THE WORK BENCH.

THE WORKPLACE

The ideal workplace (Fig. 1) will have a waterproof shed which is open on three sides. The fourth side should be arranged for laying out the tools and toolboxes. If possible there should be a lockable store for tools and materials next to the workshed. The workshed should be close to the timber pile.

THE WORK BENCH

The bench is used for supporting the workpiece while it is being marked and during the various operations of its construction. Therefore, the bench must be strong, rigid and made from good wood.

The top must be flat and it is constructed out of planks which are 5 cm thick and 30 cm wide. The length of the bench can be from 2 to 4 metres. The height of the bench can vary between 80 and 90 cm. The legs should be well braced.

There are single and double width work benches. The last type is commonly used at the building site. It must be wide enough to handle door and window frames and long enough (a full board length of 4 metres) to be used for bending concrete iron.

If the bench is built on the building site, it is not necessary to use extra timber. The timber used in the bench can be reused for some other workpiece after the carpentry work is finished.

Permanent work benches (Fig. 2) used in a workshop will be constructed slightly different from the types used at a construction site (Fig. 1). A wooden vice is used to hold the timber, with the help of a G-clamp (Fig. 2).

NOTES:

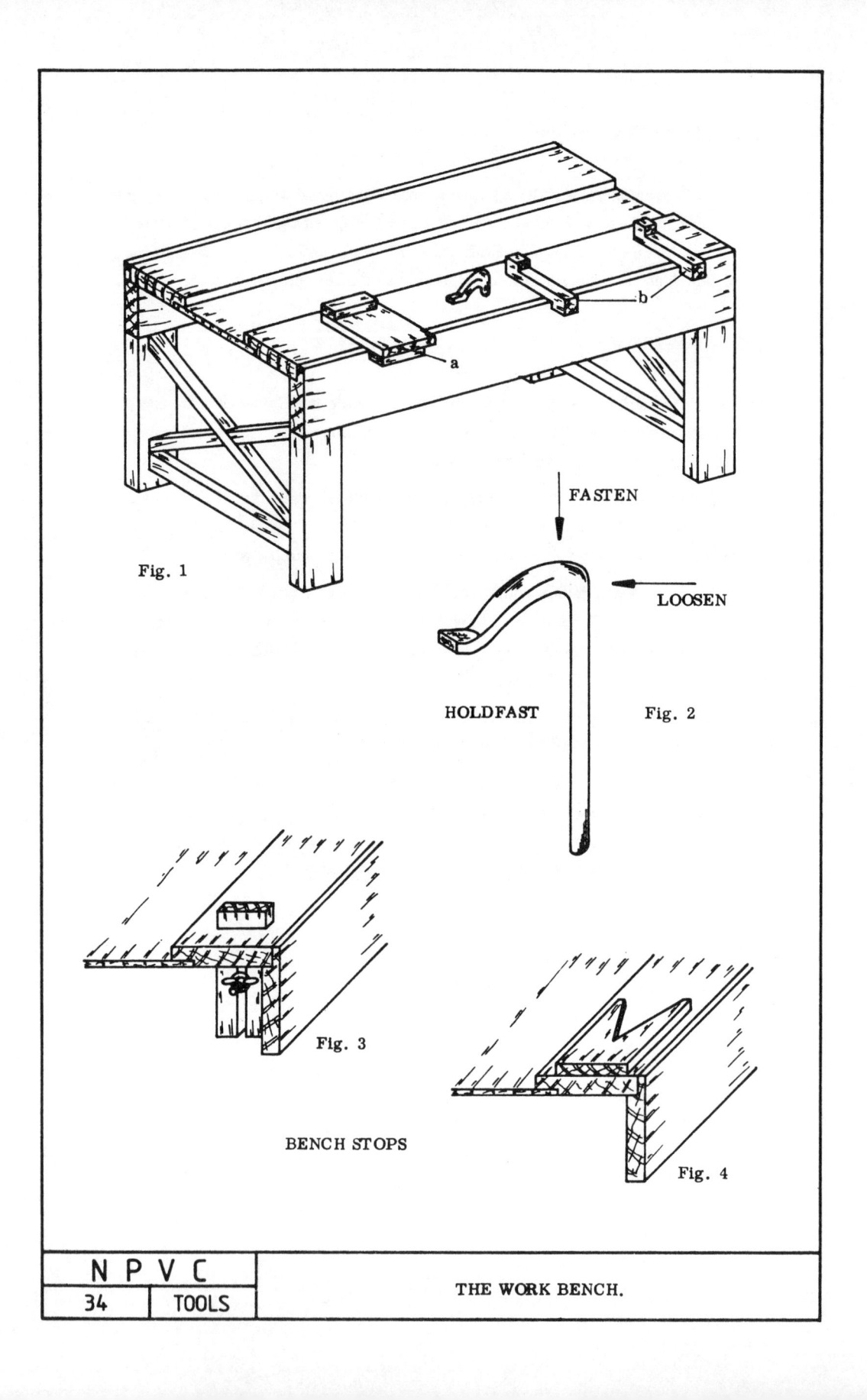

Fig. 1

FASTEN

LOOSEN

HOLDFAST   Fig. 2

Fig. 3

BENCH STOPS

Fig. 4

N P V C	
34	TOOLS

THE WORK BENCH.

BENCH HOOKS

When we have to cut a short piece of timber across the grain, we use a wide bench hook to support it. The bottom batten of the hook is held against the side of the bench and the work is pressed against the upper batten (Fig. 1a).

For long pieces of timber, we use small bench hooks for support (Fig. 1b).

Wide bench hooks measure approximately 15 x 25 cm; small ones about 5 x 25 cm.

THE HOLDFAST

A holdfast can be used to fasten wood firmly to the top of the bench. It is made tight or loosened by knocking it with a hammer (Fig. 2).

Holdfasts can be made locally from a piece of concrete iron, 2,5 cm in diameter, with a piece of a car spring welded to the top.

THE BENCH STOP

On the left side of the bench there is a bench stop made from hard wood, to support the timber during planing.

Some bench stops can be moved up or down. These are called adjustable bench stops (Fig. 3). Other bench stops are stationary (Fig. 4).

NOTES:

Fig. 1

a

b

c

Fig. 2

Fig. 3

Fig. 4

Fig. 5

X   Y

Fig. 6

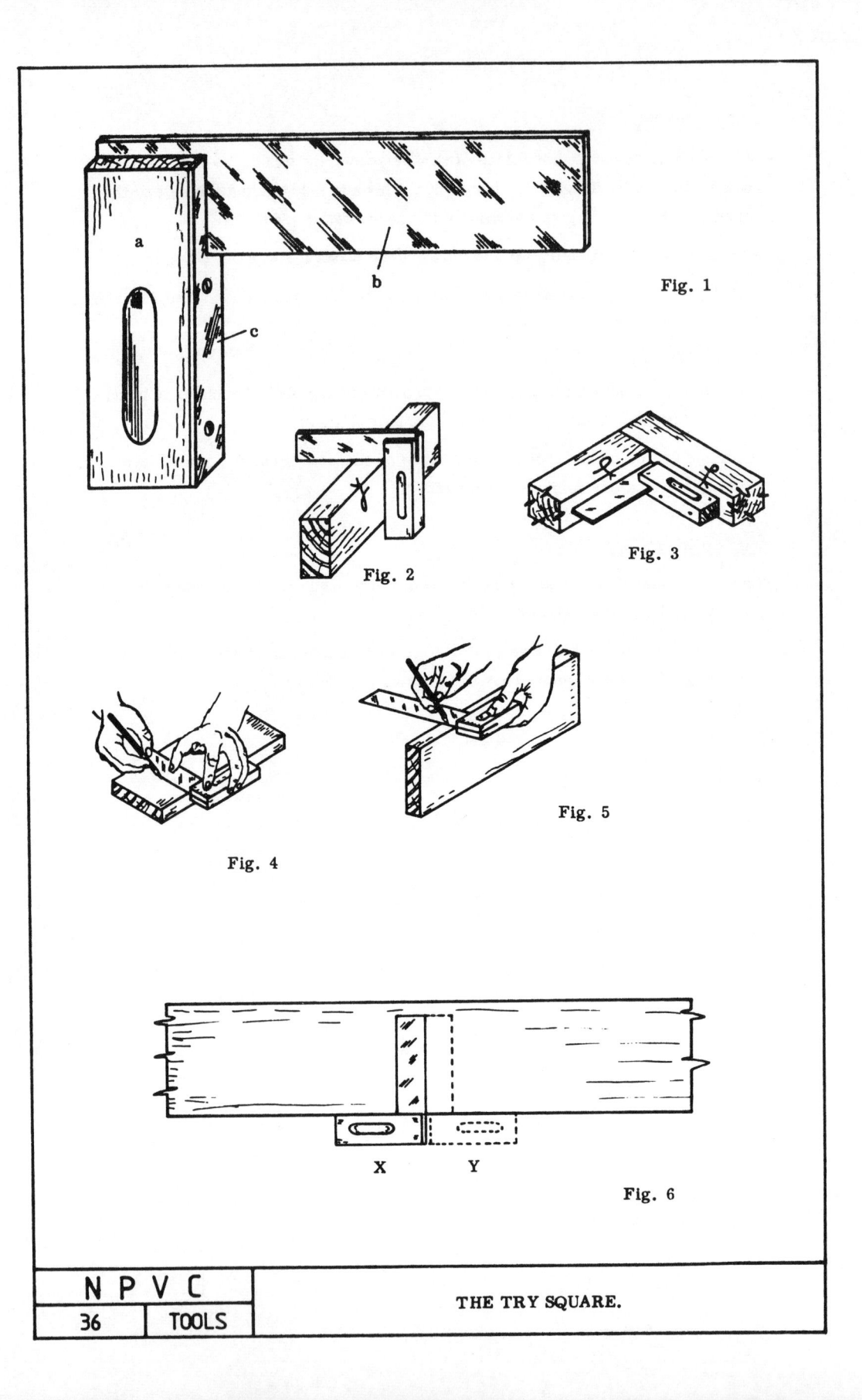

N P V C	
36	TOOLS

THE TRY SQUARE.

THE TRY SQUARE

The try square is used for marking timber, and for testing right angles to make sure that they are correct.

Its parts are (Fig. 1): the stock (a), the blade (b) and the stockface (c). The stock can be all metal or it can be made of hard wood with a brass stockface. The blade is made of steel. The angle between the stockface and blade is exactly 90 degrees.

HOW TO USE THE TRY SQUARE

- To test the angles of workpieces and boards, place it as shown in the figures on the left (Figs. 2 & 3). Always use the try square with the stock against the face edge or the face side of the workpiece when you are squaring or testing angles.

- For marking timber, press the stockface against one edge or side of the workpiece and use the blade to guide your pencil (Figs. 4 & 5).

- Keep the pencil pressed to the blade, to avoid making double lines.

- Be careful not to drop the try square or use it carelessly. Any small movement of the blade will make it inaccurate.

HOW TO TEST THE TRY SQUARE

1. Select a board with a true edge.
2. Lay the blade in position "X" and draw a line along the blade (Fig. 6).
3. Turn the square over as shown by the dotted lines (position "Y"). If the line and the blade come together exactly, the angle of 90 degrees is true.

NOTES:

Fig. 1

Fig. 2

Fig. 3

Fig. 4

Fig. 5

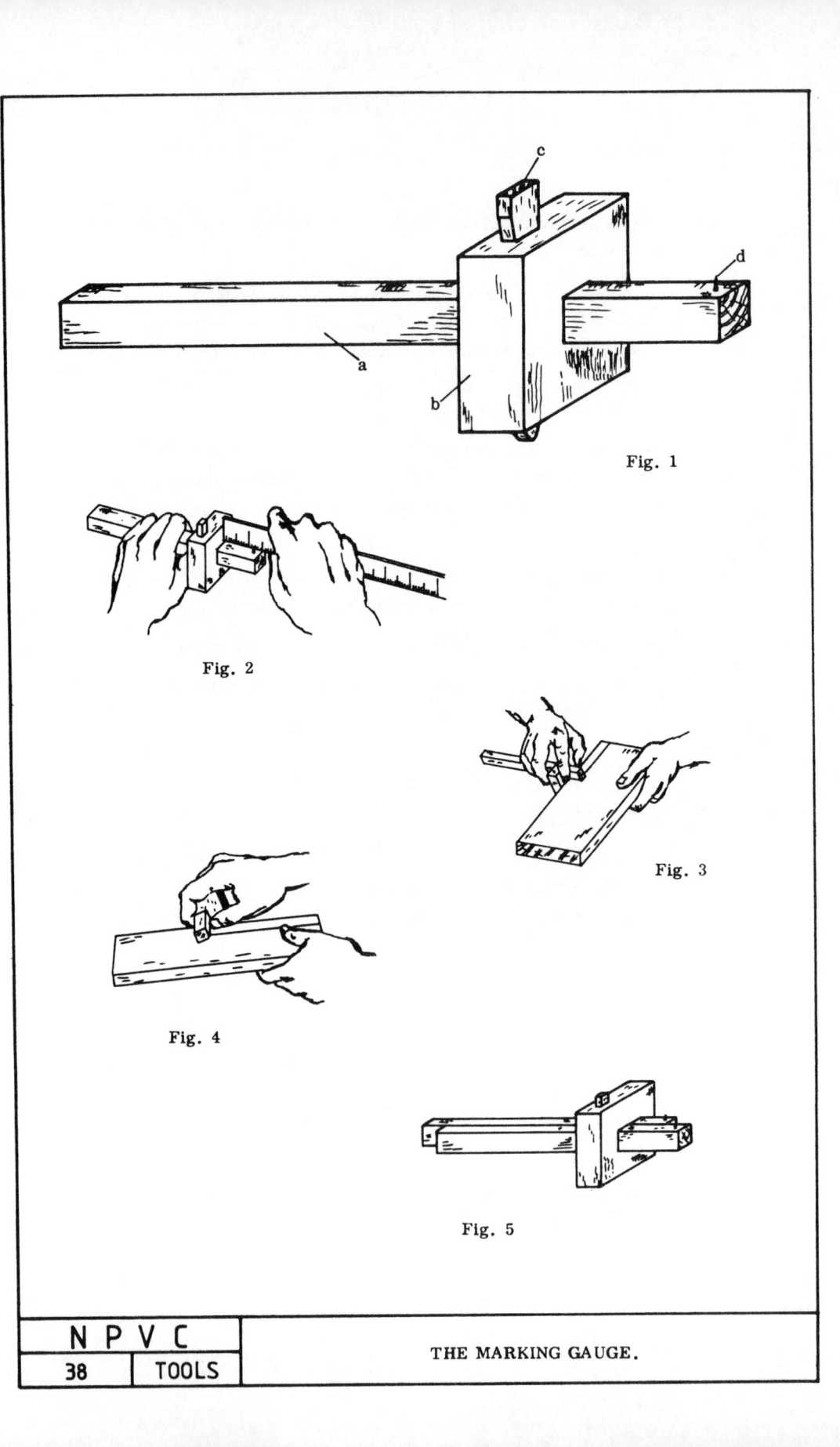

N P V C	
38	TOOLS

THE MARKING GAUGE.

THE MARKING GAUGE

The marking gauge (Fig. 1) is used to make lines on timber, parallel to the edge of the timber; that is, the lines always continue at the same distance from the edge.

The parts of the marking gauge are: the stem (a), the stock (b), the wedge (c) and the spur or pin (d). Sometimes a screw is used instead of a wedge.

HOW TO SET THE GAUGE

1. Use a rule to measure and slide the stock along the stem until the stockface is at the required distance from the pin (Fig. 2).
2. Slightly tighten the wedge or screw.
3. Check the size once more and make any necessary adjustments by tapping the end of the stem on the work bench.
4. Finally, properly tighten the wedge or screw.

HOW TO USE THE GAUGE

- Hold it in your right hand, with the face side of the stock pressed against the edge of the wood (Fig. 3).
- Keep it tilted slightly forward, so that the pin drags lightly along the wood. Don't try to make a deep mark with the pin (Fig. 4).
- Push it away from you. The pin will trace a line on the wood (Fig. 4) parallel to the edge of the wood. The stock must be held firmly against the timber edge as you move the gauge along.
- The pin may be forced out of line by grooves in the wood structure; if that occurs, mark from the other direction.
- If the gauge is hard to use at first, steady it by holding one end of the wood against the bench stop or hook.

Sometimes, marking gauges with two stems are used for marking out mortices. They are called Mortice gauges (Fig. 5).

NOTES:

Fig. 1
Fig. 2
Fig. 3
Fig. 4
Fig. 5
Fig. 6
Fig. 7

NPVC	
40	TOOLS

THE CLAWHAMMER & THE MALLET.

THE CLAWHAMMER

The clawhammer (Fig. 1) is used in light or heavy carpentry work, for driving and extracting nails. Its size, determined by the weight of the head, can be from 350 to 650 gr.

A hammer has two main parts: the head (a) and the handle (b). The head is made of steel, with a hardened face (c) and a claw (d) for extracting nails. The wooden handle is held in the eye (e) by metal wedges (f). The handle is usually made of hard wood, which absorbs shock better than metal and keeps the arm from getting tired so quickly.

HOW TO PULL NAILS

To pull a nail, slip the claws under the nailhead and pull up and back on the hammer handle (Figs. 2 & 3). When the handle reaches the vertical position and the nail is not yet all the way out, use a block of wood under the hammer (Fig. 4) to help. Pulling the handle back too far may overstrain or possibly break it.

HOW TO DRIVE NAILS

Nailing is covered in detail in the Basic Knowledge book, pages 92 to 94.

- To drive nails when you have only one hand free, hold the nail on the hammer as shown in Fig. 5 and start it in the wood with one sharp blow.

THE MALLET

The mallet is used for driving chisels, assembling joints and knocking together pieces where the hammer would damage the wood or the chisel handle. The mallet (Figs. 6 & 7) has two parts: the head (a) and the handle (b), which passes through a tapered mortice in the head.

The head can be square or round in shape. The round type can be strengthened with a metal collar (c) to prevent it from splitting. This can be made from an old shock absorber mantle.

NOTES:

THE CLAWHAMMER & THE MALLET.

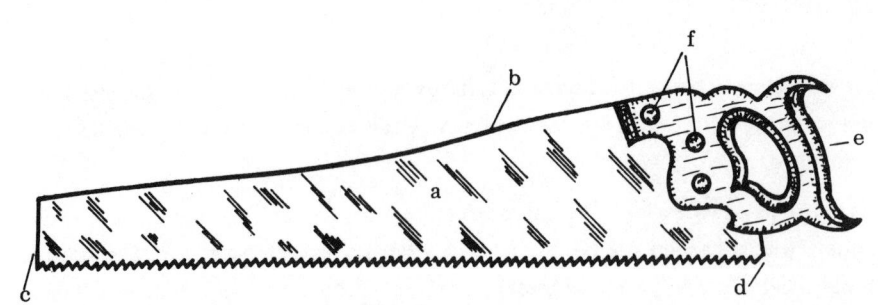

Fig. 1

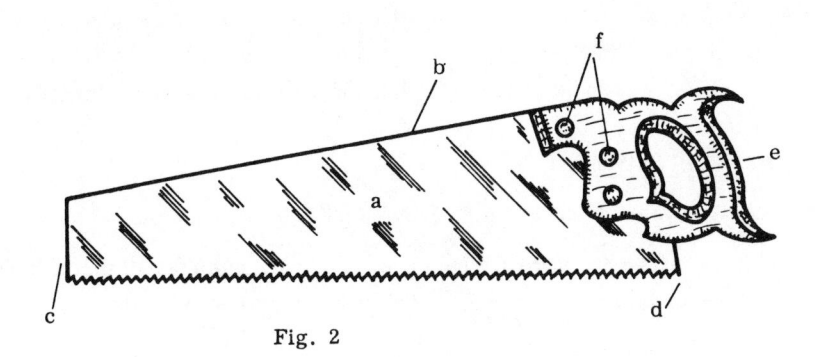

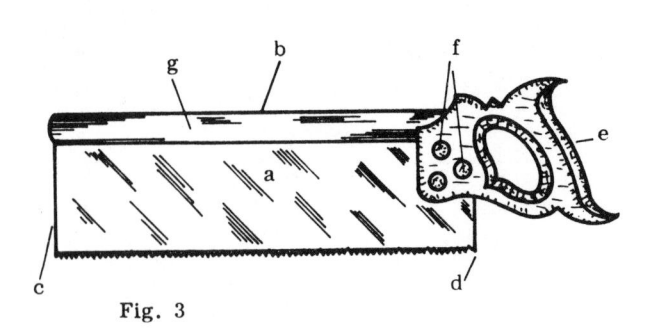

Fig. 2

Fig. 3

N P V C		HANDSAWS.
42	TOOLS	

HANDSAWS

Saws are used for cutting timber to the required size and shape with a minimum of waste in materials and labour. The principal types used in Rural Building are handsaws.

The parts of the saw are (Fig. 1): the blade (a) which has a back (b), toe (c) and heel (d), the handle (e) and the sawscrews (f) which hold the handle to the blade.

Better quality saws are taper ground, that is, they are thinner towards the back of the blade than at the cutting edge. Such saws can run (move) more freely in the kerf (the saw cut).

A good sawblade makes a clear sound when it is slightly bent and struck on the back with a fingernail. The handle should be made of good hard wood and have a comfortable grip.

Manufacturers make saws in various grades, of hard or soft steel, regular or light weight to suit any need. Depending on the kind of work we want to do, we use one of three types of handsaw: a ripsaw, a crosscut saw or a backsaw.

THE RIPSAW

The ripsaw (Fig. 1), because of the special shape of its teeth, is used only to cut with the grain of the wood. The length of the saw can vary from 66 to 71 cm. A long saw can cut faster, but it is harder to control.

THE CROSSCUT SAW

The crosscut saw (Fig. 2) is designed for cutting across the grain. Its length can vary from 51 to 66 cm.

THE BACKSAW

The backsaw (Fig. 3) is used to make fine and finished cuts. In general, the blade of the backsaw is thinner than that of a ripsaw or crosscut saw. The fold of steel that sits on the back of the blade (g) makes it stiff and it can be removed if necessary to make deep cuts.

NOTES:

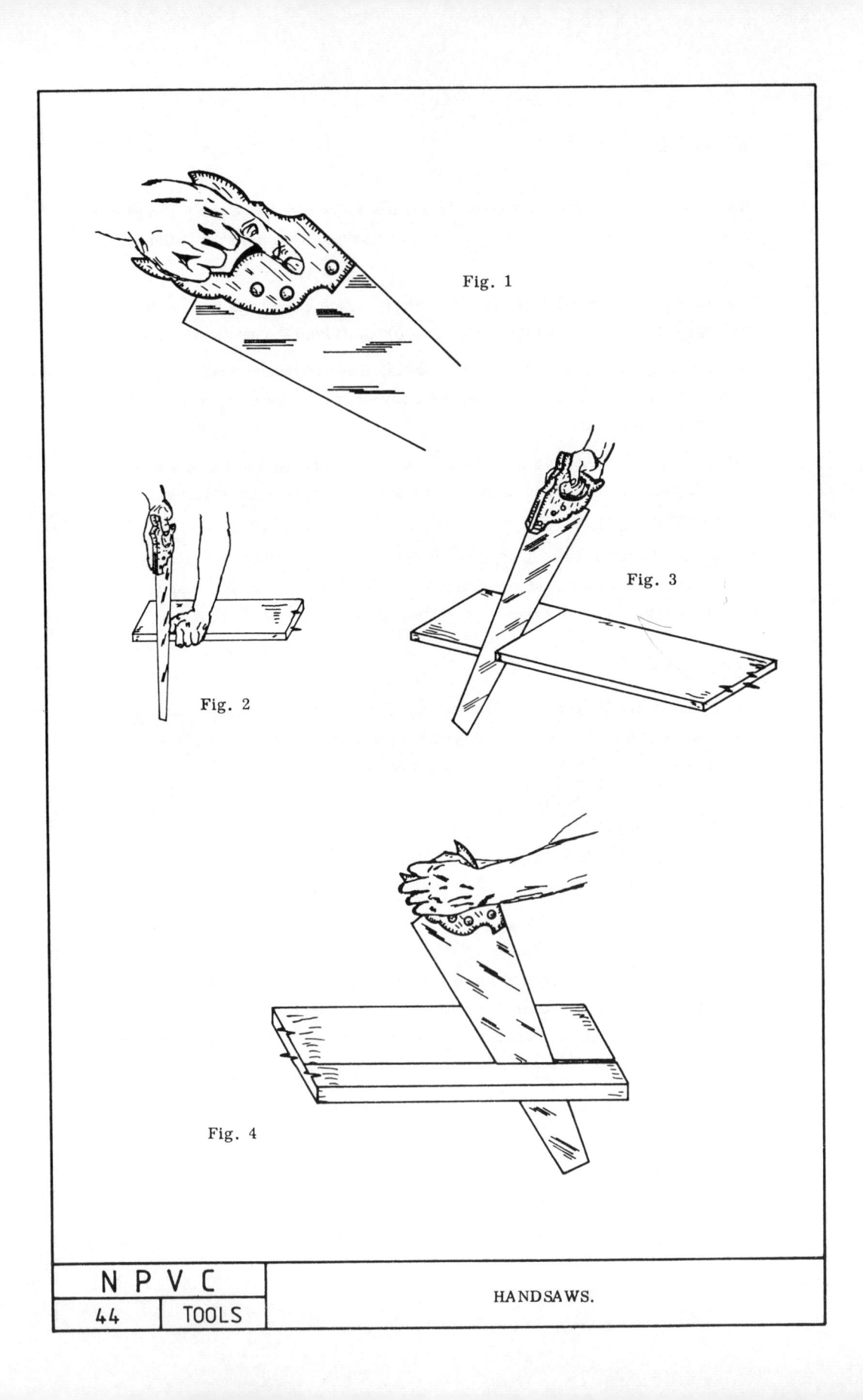

Fig. 1

Fig. 2

Fig. 3

Fig. 4

N P V C		HANDSAWS.
44	TOOLS	

HOW TO USE A HANDSAW

- Grasp the handle of the saw firmly. Your index finger should point along the blade (Fig. 1). This gives maximum control of the saw and it is the rule for holding all the different kinds of handsaws.

- To start a sawcut, grasp the far edge of the wood with your left hand, using the thumb to guide the saw while starting the cut (Fig. 2). Make two or three backstrokes, lifting the saw on the forward strokes. Draw the saw slowly and carefully, exactly along the cutting line (Fig. 3).

- After the saw is started, push it forward and pull it back, using long, easy strokes and light pressure.

- Hold the saw at an angle of about 30 degrees to the board for rough work and almost flat to the board for fine cuts.

- If the saw tends to go to one side of the line, twist the handle slightly and gently to make it come back to the line gradually, as the sawing proceeds.

HOW TO RIP BOARDS

For ripping boards (cutting with the grain), you can use different methods. The most common method used in Ghana is known as overhand ripping.

- To do overhand ripping, hold the saw in both hands as shown in Fig. 4.

When cutting timber take care that you always watch the edge of the work bench, to avoid sawing into it.

NOTES:

Fig. 1

a
g
f
e
d
b
c

Fig. 2

Fig. 3

Fig. 4

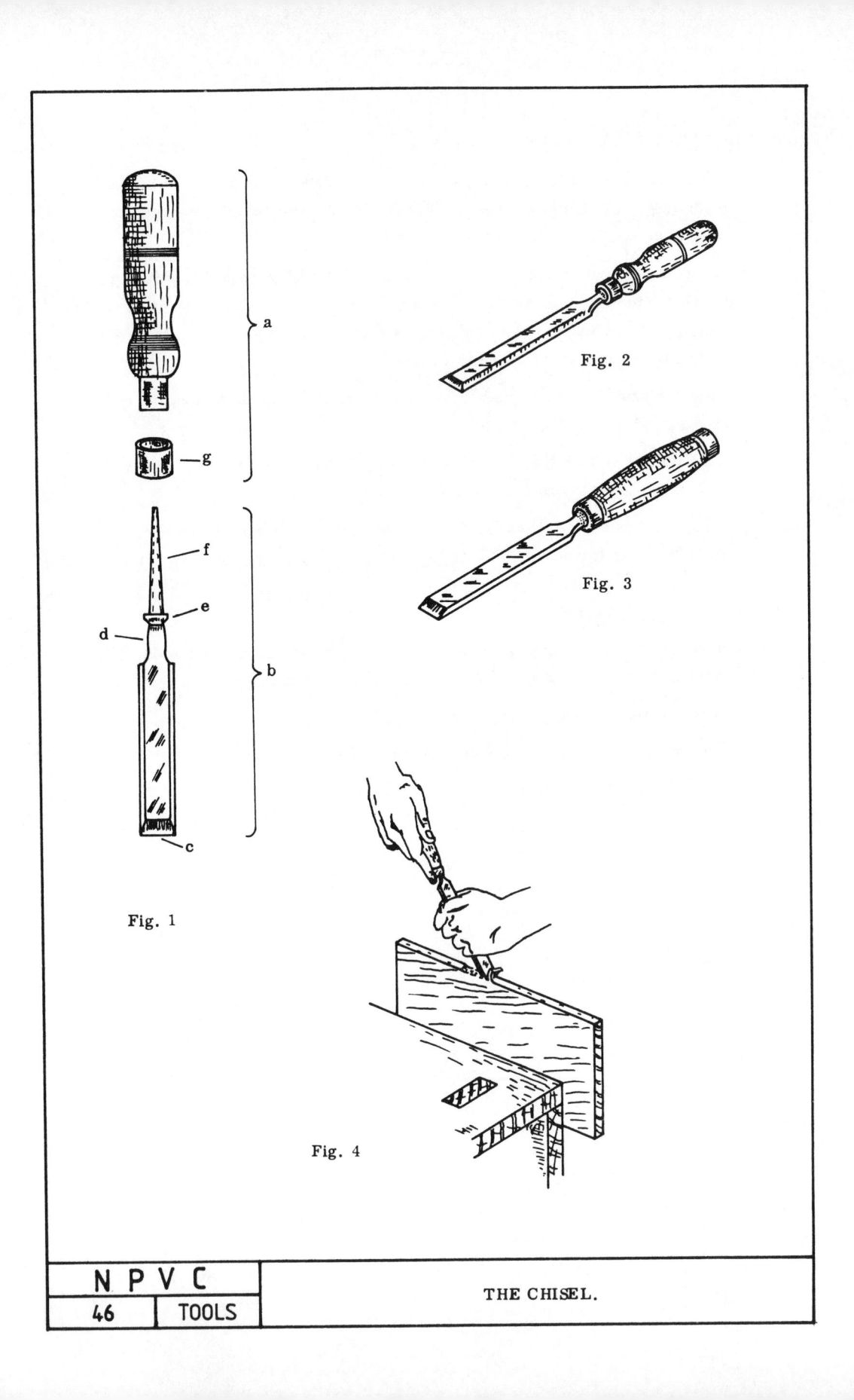

THE CHISEL.

THE CHISEL

Chisels are used for shaping wood in places where the plane cannot be used.

They have two main parts (Fig. 1): the blade (a) and the handle (b). The blade is made of steel, with a cutting edge (c) which is 3 to 32 mm wide and ground at an angle of 25 degrees. The neck (d) is the narrow part at the top of the blade. The shoulder above the neck (e) is to prevent the blade from being driven too far into the handle and splitting it. The tang (f) is the end of the blade which fits into the handle and holds the two parts together.

The ferrule (g) at the bottom of the handle keeps the wood from splitting where the blade enters the handle. The handle itself is made from hard wood or plastic and it is slightly rounded on top to prevent splitting.

For the various tasks of carpentry work in Rural Building, there are different kinds of chisels. The two most common ones are the firmer chisel and the mortice chisel.

THE FIRMER CHISEL

Most of the firmer chisels (Fig. 2) are the type with a bevelled edge. The firmer chisel is normally driven into the wood by hand and is used only for light cutting and shaping work.

THE MORTICE CHISEL

The blade of the mortice chisel (Fig. 3) is thicker and stronger than that of the firmer chisel, as it is used for heavy work. This chisel is driven into the wood with a mallet (never with a steel hammer), so it is usually fitted with two ferrules, at the top and bottom of the handle, to prevent splitting.

NOTES:

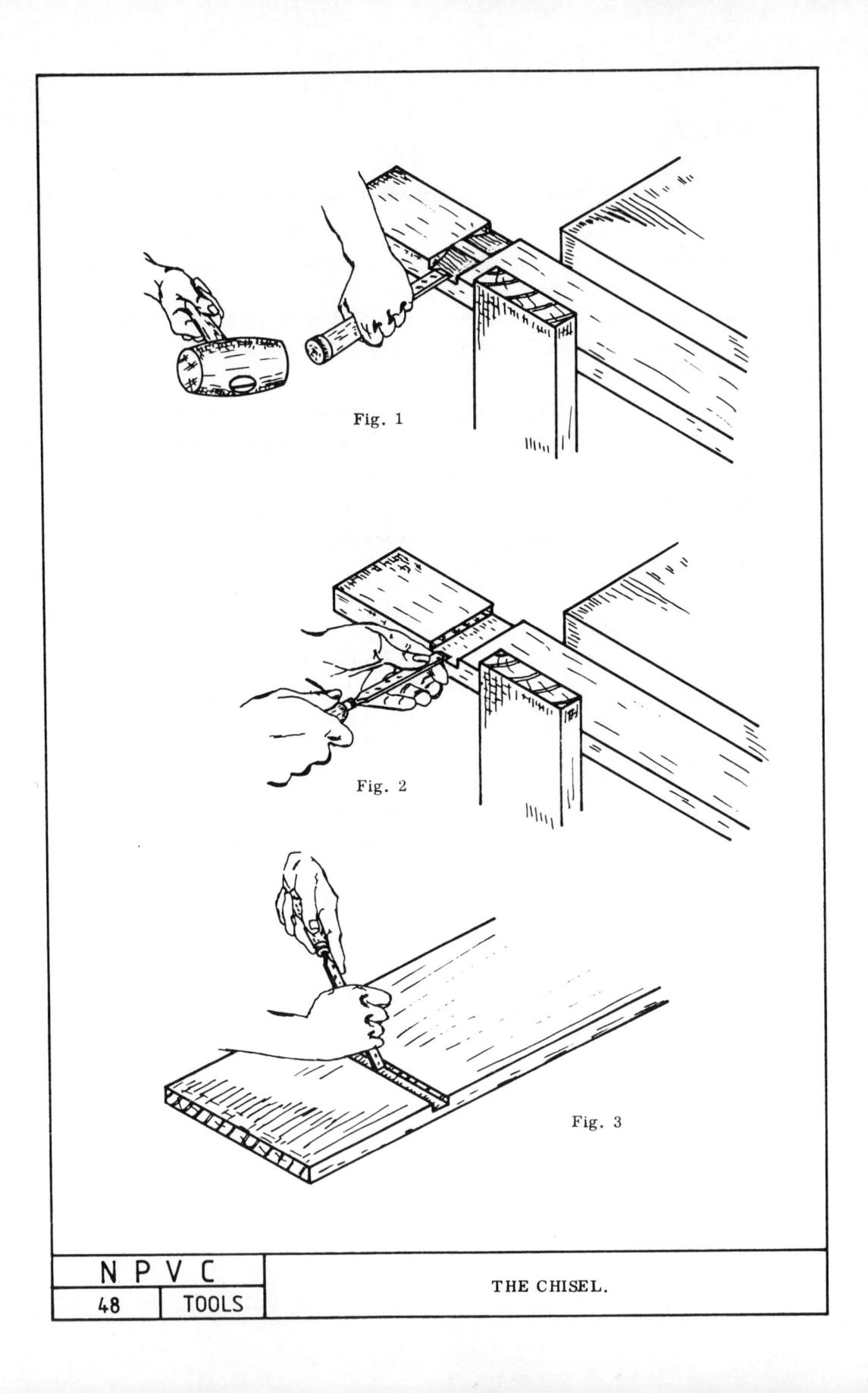

Fig. 1

Fig. 2

Fig. 3

N P V C		
48	TOOLS	THE CHISEL.

HOW TO USE THE CHISEL

To do good work, you need a sharp chisel. A dull chisel is hard to force through the wood and it is also hard to guide and control, making the resulting work rough and inaccurate. The time you use to stop and sharpen a dull chisel will soon be regained by better and faster work.

To prevent dulling the chisel, do not allow the cutting edge to touch other tools or the bench top. Always lay the chisel on the bench with the bevel side down.

CHISELLING WITH THE GRAIN

When chiselling with the grain, observe the following points (see Fig. 4 on the previous page):

- Always work with the grain, to avoid splintering or splitting the wood.
- Fasten the work securely so your hands are both free for the chisel.
- Always push the chisel away from yourself, keeping both hands behind the cutting edge.
- Use your left hand to guide the chisel, and your right hand to push on the handle.

CHISELLING ACROSS A BOARD

When chiselling across the grain, observe the following:

- Grasp the blade of the chisel between the thumb and forefinger of your left hand, to guide it and act as a brake, while the pushing is done with your right hand (Fig. 2).
- Cut with the bevel side up, raising the handle just enough to make the chisel cut. For heavier chiselling and for rough cuts, the mallet may be used as is shown in Fig. 1.
- When chiselling across wide boards, where the chisel cannot reach to the center of the board, work with the bevel side down (Fig. 3).

NOTES:

THE CHISEL.

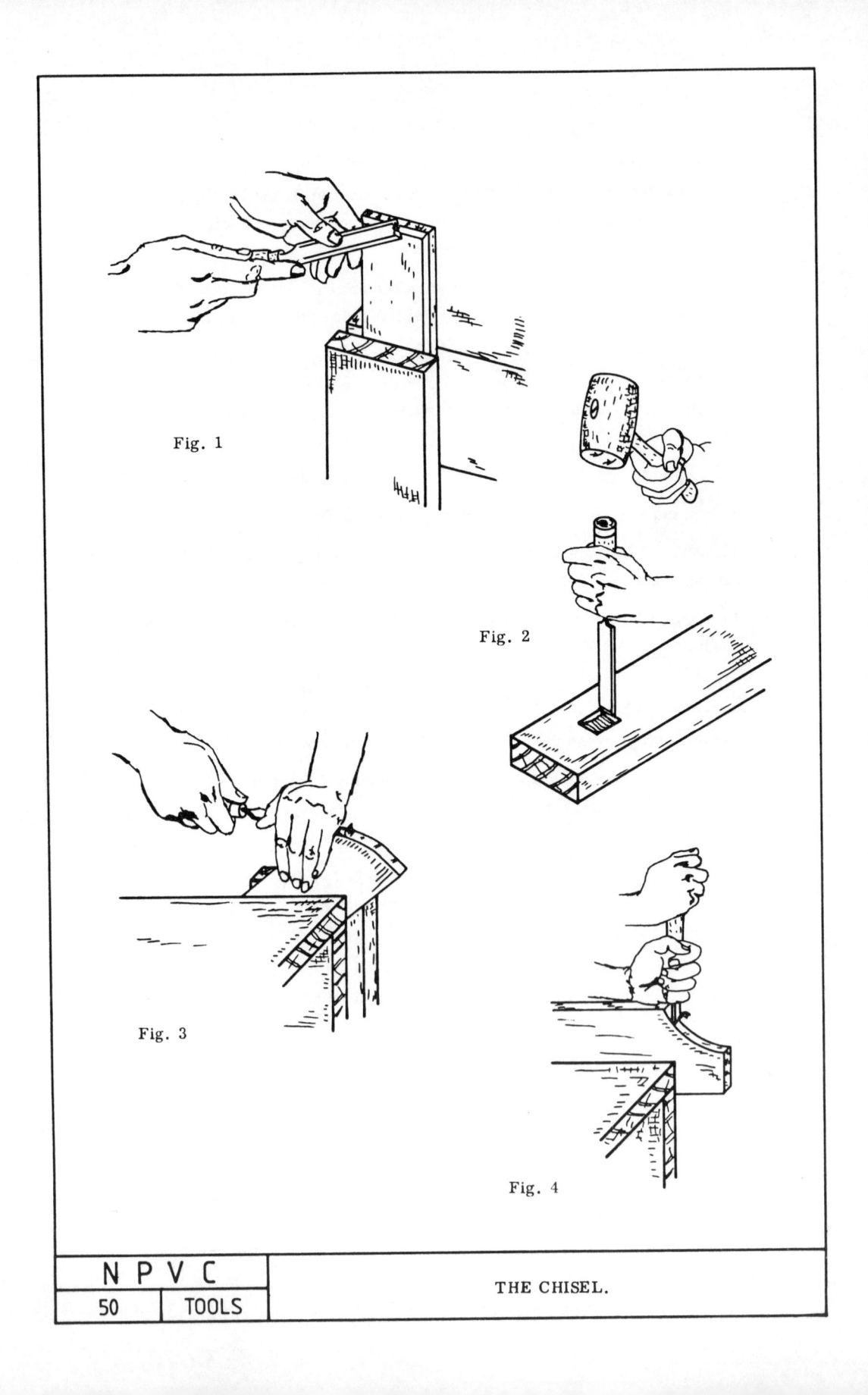

Fig. 1

Fig. 2

Fig. 3

Fig. 4

N P V C		
50	TOOLS	THE CHISEL.

USING THE MALLET

Use a mallet to drive the chisel when force is required to make deep, rough cuts. Never use a steel hammer, it will soon damage the chisel handle.

- Make a series of light taps with the mallet instead of heavy blows, as light taps will give you better control.

CHAMFERING

The chisel may be used to make chamfers or bevels, with or across the grain.

- Keep the bevel up. As you push the chisel forward, move the handle slightly from side to side, so that the cutting edge works obliquely (at an angle) (Fig. 1).

- Prevent splintering in cutting end chamfers or bevels by working part way from one edge and part way from the other.

MORTICING

Morticing is done as shown in Fig. 2 on the left page.

CUTTING CURVES WITH THE CHISEL

Convex curves can be shaped with the chisel as shown in Fig. 3.

- Use the chisel with the bevel side up. Hold and guide it with your left hand while you push it forward with your right hand.

Concave or inside curves may also be finished with the chisel, as in Fig. 4.

- Use the chisel with the bevel side down.

- Guide it with your left hand, while your right hand pushes down and pulls backward at the same time.

NOTES:

THE CHISEL.

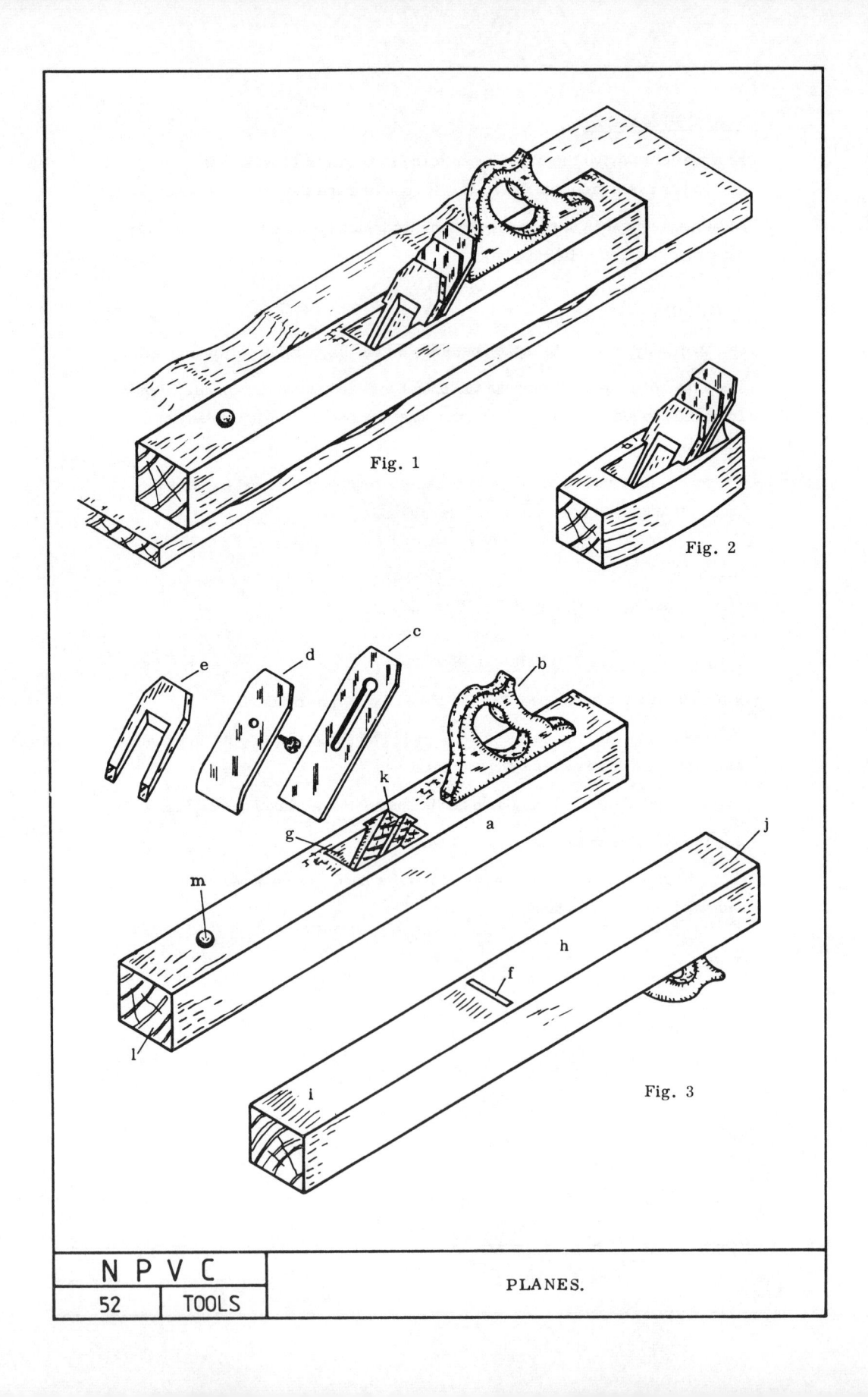

Fig. 1

Fig. 2

Fig. 3

N P V C	
52 TOOLS	PLANES.

PLANES

Together with the hammer, saw, try square and chisel, the plane is one of the principal tools used for carpentry work.

When timber comes from the sawmill, it is rough from the saw. Before rough-sawn timber or boards can be used for any finished work, they must be prepared so that all their sides and edges are square, flat and smooth. This preparation is done by planing them first and later smoothing them.

The tool used for the rough part of this work is called a jack plane (Fig. 1). It is used for planing the wood true and to reduce it to the correct size. The jack plane is used for general planing purposes, which is why it is called "jack", meaning generally useful.

After the board has been planed to a true surface, we use the smoothing plane (Fig. 2) to remove the rough marks of the jack plane, making a smooth surface on the board.

Other planes, used for special purposes, are discussed later in this book.

THE JACK PLANE

The jack plane (Fig. 3) consists of five main parts: the stock or body (a), the handle (b), the cutting iron (c), the cap iron (d) and the wedge (e).

The stock of the plane is made of hard wood which wears uniformly (doesn't get more worn out in some areas than in others), is tough, straight grained and keeps its shape. It is about 40 cm long. This long length lets the plane go over the low spots in the wood without cutting (Fig. 1), while it removes the high spots.

The stock holds the cutting iron, which takes off the shavings. The cutting iron sticks out through the mouth (f) on the bottom of the stock and the shavings leave the plane through the escapement (g) on top of the stock.

The bottom of the stock, which rubs along the wood during planing, is called the sole (h). The front part of the sole is called the toe (i) and the back part is the heel (j).

The cutting iron rests on the bed (k), at a 45 degree angle to the sole.

The front end of the stock is called the nose (l). To remove the cutting iron or to reduce the cut, the top of the nose is struck sharply with a hammer. In order to prevent the wood from being bruised, some planes have a small piece of metal let into the nose, to take the blows of the hammer. This is called the striking button (Fig. 3, m, previous page).

The handle is morticed into the stock behind the escapement. It should have a comfortable grip to protect the hand.

The cutting iron is made out of steel. It has a slot in the centre which ends in a screw hole. The cutting edge is ground at an angle of 25 to 30 degrees (see Maintenance of Tools, pages 92 to 99). If there is any imperfection in the cutting edge you will see a mark from it on the wood after planing.

The cap iron is made from mild steel and is secured to the cutting iron by a cheesehead screw (holding screw) which passes through the slot in the cutting iron. The cap iron helps the cutting edge to plane smoothly, by breaking up the shavings before they split ahead and tear up the fibres of the wood.

It is essential that the back of the cap iron should bed perfectly on the face of the cutting iron when they are fitted together, for even the slightest gap between the two will allow a shaving to enter and the mouth of the plane will immediately become blocked (see Maintenance of Tools, page 104).

The wedge is made of hard wood and its function is to hold the irons in place. It fits into special grooves in the sides of the escapement.

It is a good idea to keep the wooden parts of the plane in good condition by occasionally rubbing some vegetable oil, like groundnut oil (not machine oil) on them.

NOTES:

THE SMOOTHING PLANE

The smoothing plane is constructed similarly to the jack plane. It is used to finish the surface of the wood.

The smoothing plane is about 20 cm long, so it is the smallest plane that a builder uses (Fig. 2, page 54).

The smoothing plane is not used to remove large quantities of wood or to plane the wood flat and true. It is set to remove only rather fine shavings, so we use it only to make a smooth surface on the wood.

The smoothing plane has a handguard rather than a handle like the jack plane and it has the striking button on the back end of the plane instead of on the nose.

NOTES:

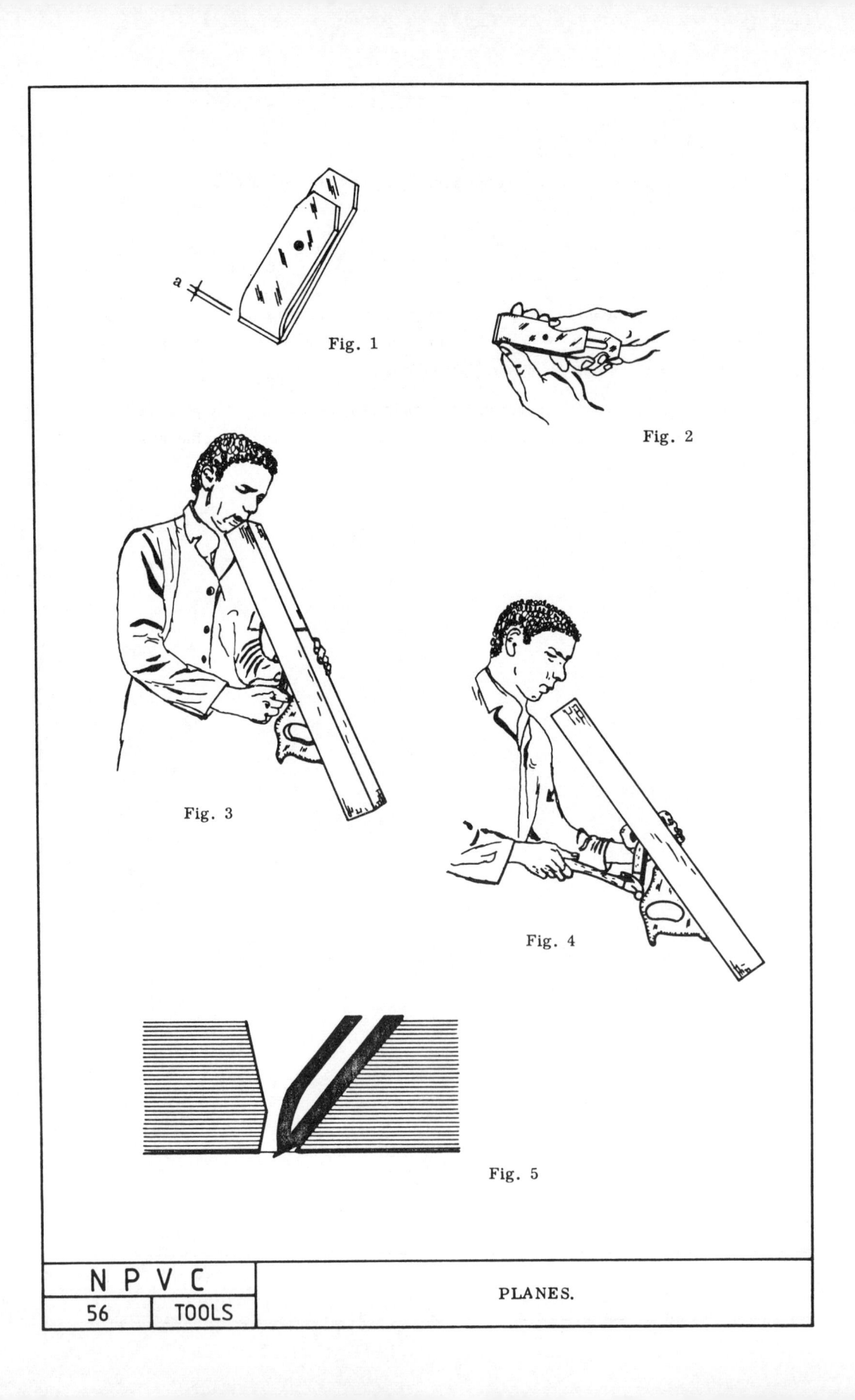

Fig. 1

Fig. 2

Fig. 3

Fig. 4

Fig. 5

N P V C		PLANES.
56	TOOLS	

HOW TO SET THE CAP IRON

The distance from the edge of the cap iron to the cutting edge is called the set (Fig. 1, a).

- The set for the jack plane should be:
  2 mm, when planing hard woods;
  3 mm, when planing soft woods.

- The set for the smoothing plane should be:
  1 mm or less.

In order to set the cap iron, first loosen the cheesehead screw which holds the two irons together; until the cap iron can move. Hold the irons as shown in Fig. 2 and be very careful that the cap iron does not touch the cutting edge as this will make it dull. Adjust the irons to the proper set and tighten the screw.

HOW TO SET THE CUTTING IRON

1. Hold the plane in your left hand and look along the sole while putting the irons in it (Fig. 3). Fix the wedge so that they are held in place lightly. Adjust them to roughly the correct position.

2. Adjust the cutting iron exactly by tapping with the clawhammer either on top of the iron (Fig. 4), so that it comes out more and takes off thicker shavings; or on the striking button to get thinner shavings.

3. After every adjustment, tighten the wedge slightly. Don't hit it too hard or you will damage the wedge.

The cutting iron should project as shown in Fig. 5 on the left page. It is important that the cutting iron projects evenly. For rough work, you will want the cutting iron to project more so as to take off thicker shavings. For fine work, where only a little wood will be planed off, the cutting iron should project less.

- After you finish using the plane, knock the cutting iron back so that it doesn't project out of the sole at all. This is to prevent damage to the cutting edge when the tool is not in use.

NOTES:

Fig. 1

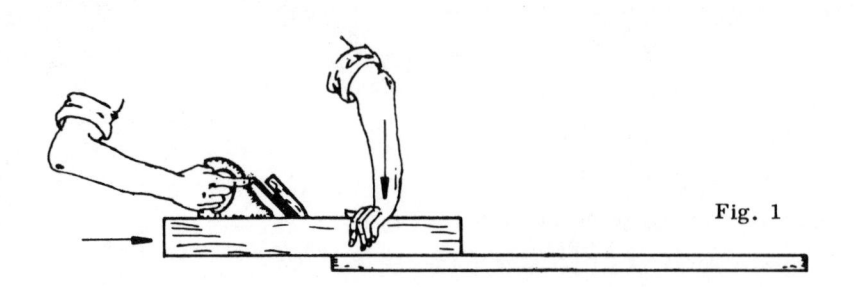

PRESS, PUSH

Fig. 2

AND LIFT

Fig. 3

N P V C		
58	TOOLS	PLANES.

PLANING

- When you use the plane, try to put pressure on it in such a way that the part of the sole which is in contact with the wood is pressed firmly to the wood surface. At the start of the stroke when the cutting iron does not yet touch the wood, put pressure on the front part of the plane (Fig. 1). When the cutting iron comes to the end of the board, press down more on the heel of the plane (Fig. 2).
- Guide the plane when you plane on edges, by curling your fingers under the plane so that they contact the board (Fig. 3).
- To obtain a good surface, always plane with the grain. If the wood is cross-grained, it is best to hold the plane at an angle to the direction of the stroke.

STUFFING

Sometimes the plane won't take off shavings anymore, but just slips over the wood without cutting it. This happens when shavings have blocked up the plane: the plane is stuffed with shavings.

- The plane may stuff when the cap iron is not fitted well to the cutting iron. Shavings can enter the gap between the irons and block the mouth of the plane (see Maintenance of Tools, page 104).
- Another cause of stuffing may be an incorrectly made plane. If you make your plane by hand, be sure to make the mouth and the escapement large enough. The mouth must not be too big however, because then it will not give a good surface when it is used to plane.
- Make sure that the ends of the wedge are not projecting over the cap iron and blocking the shavings.

NOTES:

Fig. 1

SHARPENING STONE

Fig. 2

SHARP-EDGED GRAINS

BINDING MATERIAL

ENLARGED DETAIL

WOODEN BOX

PANEL PIN

Fig. 3

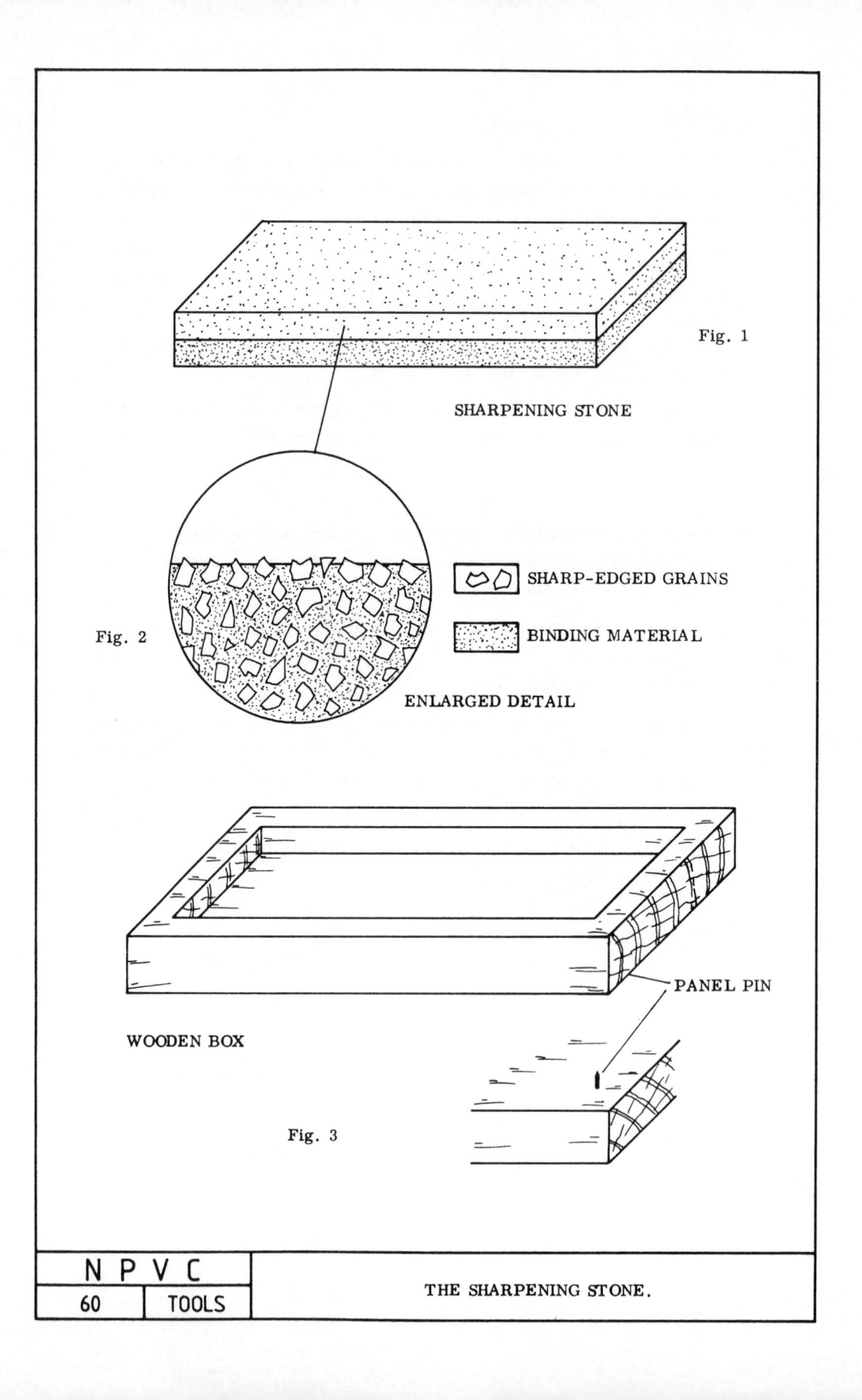

N P V C		
60	TOOLS	THE SHARPENING STONE.

THE SHARPENING STONE

The sharp edges of plane irons and chisels are made with sharpening stones. The stone acts like a file to wear away the portion of the tool that is rubbed on it.

There are natural and artificial (man-made) stones. Nowadays we use mostly the artificial ones. The grit (rough surface) of the stone consists of hard, sharp-edged grains, embedded in a binding material (Fig. 2). The bigger the grains, the coarser the stone. Stones can be coarse, medium, or fine; in Rural Building we use a combination stone, one side of which is coarse and the other fine (Fig. 1). The coarse side cuts quickly, but does not give a very keen (sharp) edge. The fine grit cuts slowly and gives a keen edge.

Take very good care of your sharpening stone. House it in a solid wooden box. On each corner of the box, drive a panel pin almost home and file the heads to sharp points (Fig. 3). The pins will anchor the box and the stone to the bench when you are sharpening tools.

HOW TO USE THE STONE

- Before using the stone, soak it in water for a few minutes. Use water during the sharpening to wash away the metal particles. After use, clean the stone with water. The water keeps the stone from glazing (becoming clogged with small metal particles between the grains, which makes it smoother and less efficient to use).

- Try to keep the stone worn down evenly and flat by using the entire surface, not rubbing just in the centre.

- The end nearest you is less easy to use, so turn the stone around occasionally to let both ends wear down equally.

- If a stone is worn hollow, it can be made flat again by rubbing it on a flat stone or cement surface, using sand and water to grind it.

NOTES:

Fig. 1

Fig. 2

Fig. 3

Fig. 4

Fig. 5

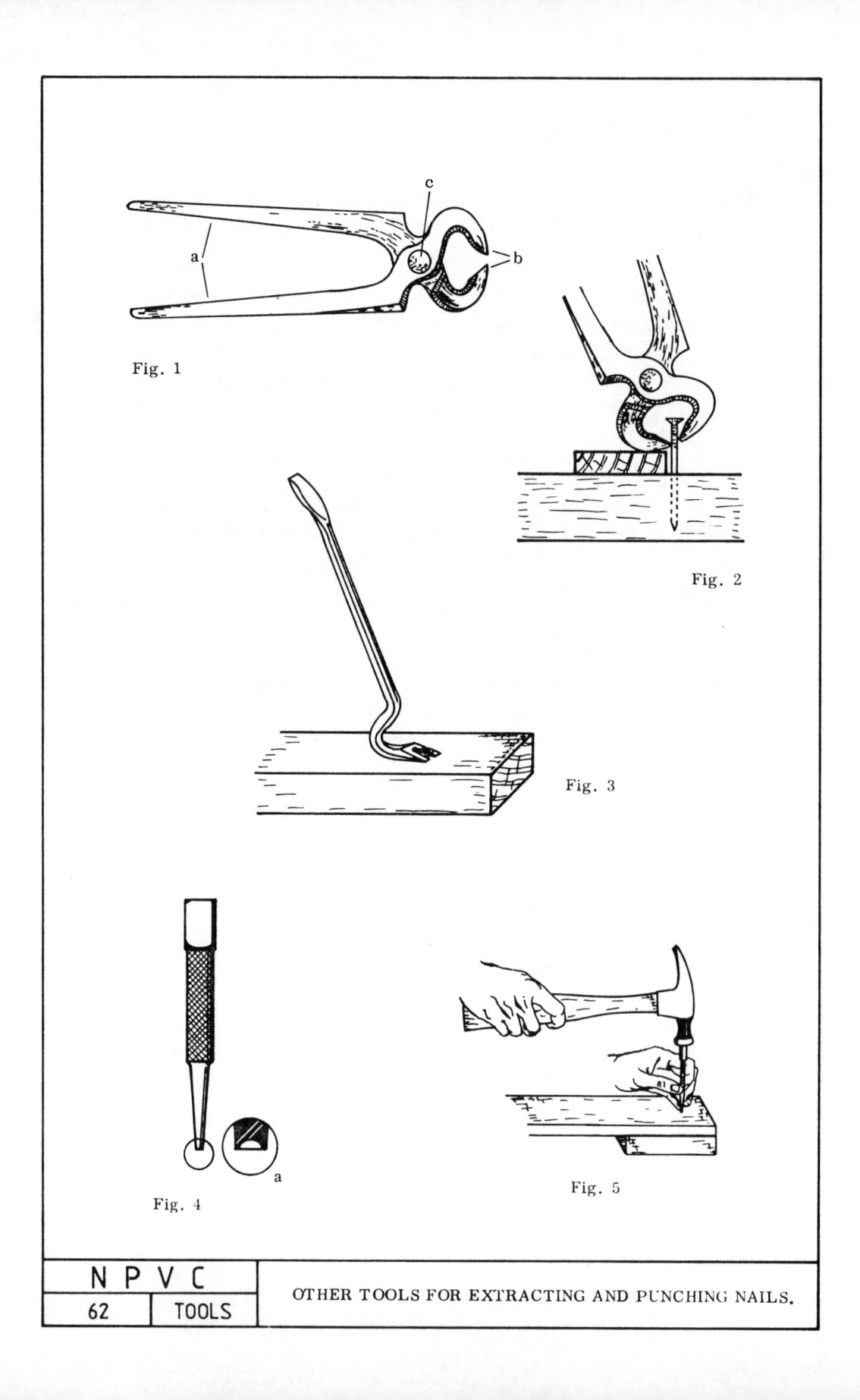

N P V C	
62	TOOLS

OTHER TOOLS FOR EXTRACTING AND PUNCHING NAILS.

OTHER TOOLS FOR EXTRACTING AND PUNCHING NAILS

PINCERS

Pincers (Fig. 1) are used chiefly for extracting nails which have become bent in driving. When using the pincers, protect the wood surface from bruising by using a small piece of wood underneath (Fig. 2).

Pincers have three main parts: the arms (a), made of steel, the jaws (b), made of hardened steel and sharpened to grip nails etc. and the rivet (c) which connects the two arms.

THE CROWBAR

This is an iron bar with a forged end (Fig. 3) used for pulling big nails out of timber and as a lever to move heavy objects. Other uses are: to open crates, or to loosen boards of concrete forms.

When a lot of force is needed to get out a nail, you use a crowbar so as not to break the handle of your hammer.

For lifting very heavy objects, an iron bar should be used in place of the smaller crowbar.

THE NAIL PUNCH

The nail punch (Fig. 4) is used along with the hammer to drive the heads of nails below the surface of the wood and to clench nails that go through to the other side of the board and stick out.

Nail punches are generally cylindrical in shape, with concave points to keep the punch from slipping off the nailhead (Fig. 4a). If no punch is available, a large blunt nail can be used instead.

- To punch a nail under the surface of the wood, hold the punch in place on the nailhead, steadying your hand on the board (Fig. 5), and hit it with the hammer.

NOTES:

SWIVEL SHOE

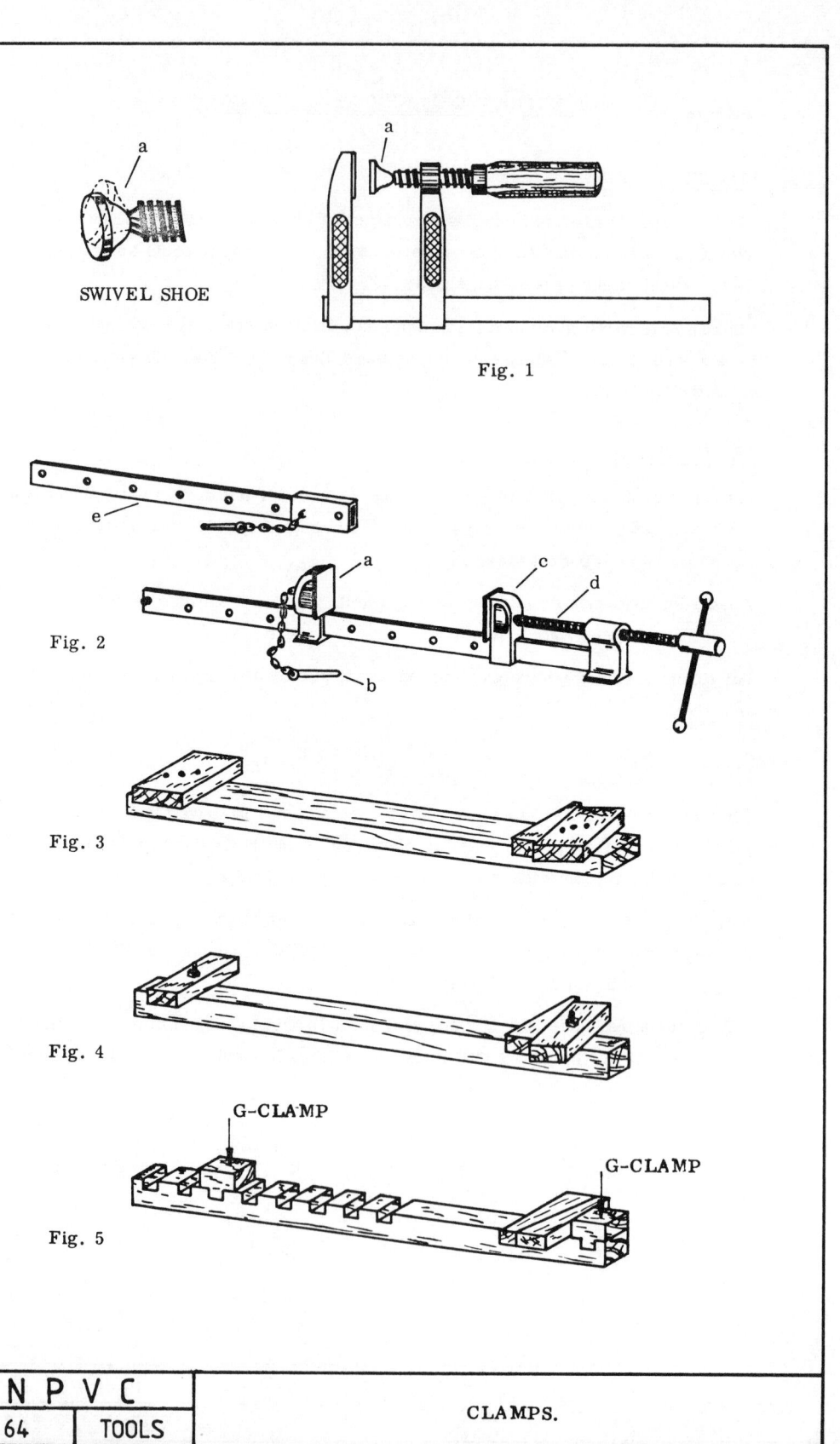

Fig. 1

Fig. 2

Fig. 3

Fig. 4

G-CLAMP

G-CLAMP

Fig. 5

NPVC 64 TOOLS CLAMPS.

CLAMPS

THE G-CLAMP

The G-clamp (Fig. 1) is used to hold the job under control while it is worked upon. It is used in the workshop as well as on the building site and can be used for holding small pieces together while they are being glued.

Except for the handle, which is made of wood, the parts are all steel. The shoe (a) is set on a swivel, which allows it to move and adapt to the surface of the job. These clamps are available in different sizes, up to 240 cm.

THE METAL SASH CLAMP

Sash clamps (Fig. 2) are used for pushing together joints and holding parts for glueing or nailing, or holding wood when making rebates.

The adjustable shoe (a) is fixed on the bar by the pin (b). The clamping shoe (c) is tightened against the job by the screw (d).

The length of the clamps can be between 100 and 200 cm, and they can be made longer by an extension bar (e).

- If more than one clamp is used on a piece, take care that they are set in line, so that they don't twist the piece.

THE WOODEN SASH CLAMP

Homemade clamps (Figs. 3, 4 & 5) can serve the same purposes as a metal one. The job is tightened by means of a wedge.

For quick jobs, a clamp can be made as shown in Fig. 3. Here the different members are nailed together according to the required size of the sash clamp.

If a more permanent, adjustable clamp is required, it can be made as shown in Figs. 4 or 5.

NOTES:

Fig. 1

Fig. 2

SWEEP

Fig. 3

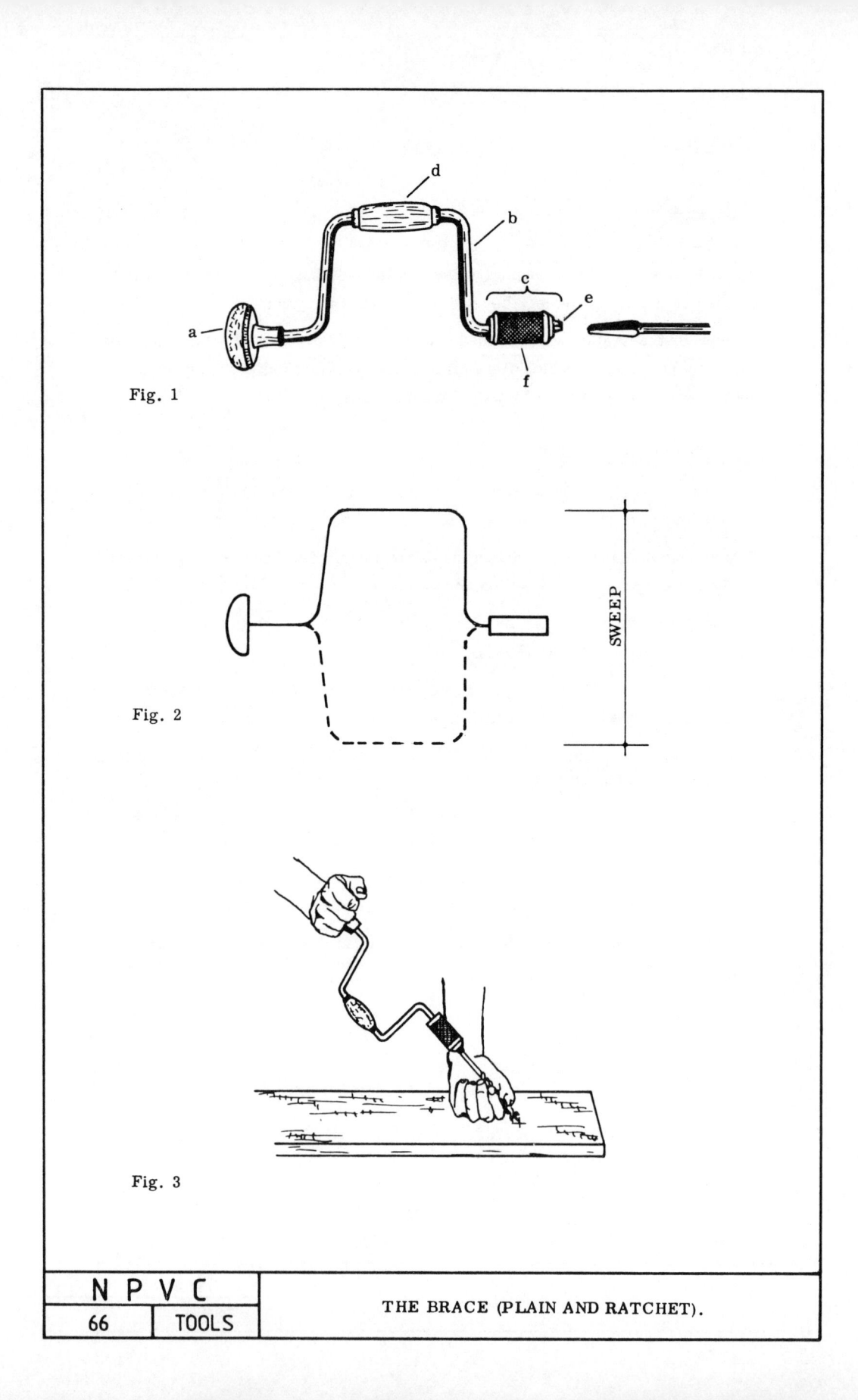

N P V C	
66	TOOLS

THE BRACE (PLAIN AND RATCHET).

THE BRACE (PLAIN AND RATCHET)

This is a cranked tool used to turn drill bits and countersinks, thus making holes in wood.

Its parts are (Fig. 1): the head (a), the crank or bow (b), and the chuck (c).

The head is a hard wooden or plastic knob, which is fixed to the bow and turns on a ball bearing.

The crank, which is formed by the rectangular bend, can be 10 to 20 cm in width giving a sweep of 20 to 40 cm, which determines the size of the brace (Fig. 2).

The wider the brace, the more force you can apply with it, but also, the distance your arm must travel is longer and more tiring and the bit may be broken more easily.

A handle (d) is attached to the crank by steel collars. It should revolve freely.

The chuck holds the various bits (the drilling parts). The bits are gripped by jaws (e). There are different kinds of jaws. The square shank of the bit fits into the square opening of the chuck. When the socket (f) is turned, it tightens the jaws over the shank of the bit.

There is a more advanced type of brace available, which is the ratchet brace. The ratchet makes it possible to use the brace in places where it is impossible to make a complete turn of the crank with an ordinary brace.

HOW TO USE THE BRACE

1. For accurate boring, first mark the location of the centre of the hole with two lines crossing each other, or by making a small hole with a sharp tool (Fig. 3).

2. With the knuckles of one hand down against the board, guide the point of the bit carefully into place, while with your other hand you exert a slight pressure on the head of the brace (Fig. 3).

3. As the bit starts boring, be careful to keep it perpendicular to the surface (unless you want the hole to be bored at an angle).

NOTES:

Fig. 1

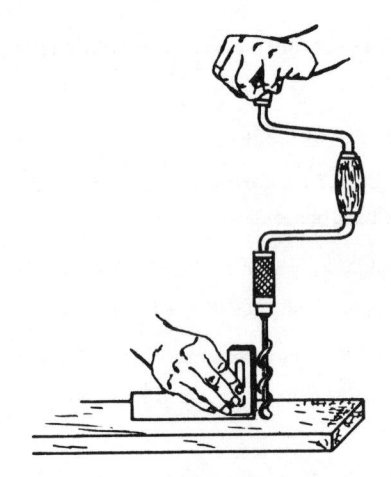

Fig. 2

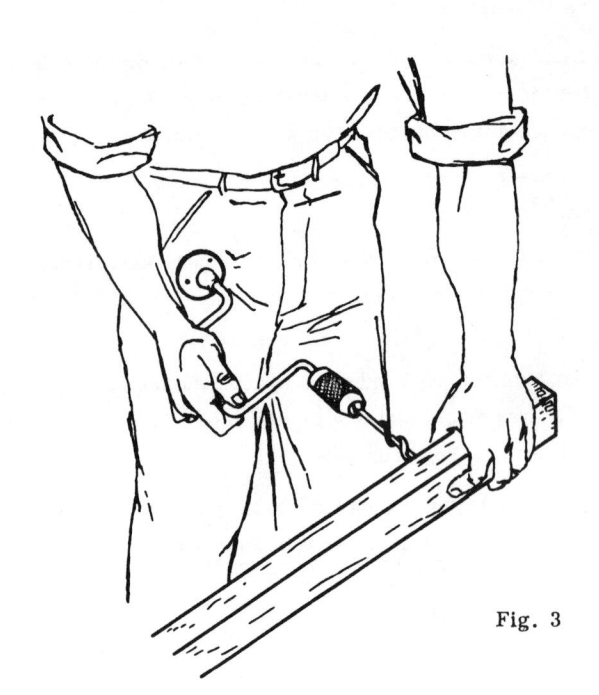

Fig. 3

THE BRACE (PLAIN AND RATCHET).

It is fairly easy to tell if the brace is leaning to the right or left, but difficult to know if it is leaning away from you or towards you.

- For jobs where this is important, it is best to ask another person to stand to one side and tell you whether you are holding the brace upright.

- Another way to see whether the brace is boring square to the surface is to step back a little, steadying the brace with one hand, and sight. Then move around and sight in the other direction, at about right angles to the first sighting (Fig. 1).

- A square can be used to check if the bit is boring straight (Fig. 2).

- If a hole is to be bored completely through a board, bore until the point of the bit can be felt on the other side (Fig. 3), then turn the board over and bore from the other side. This prevents splintering around the edge of the hole where the drill comes out.

- You can also prevent splintering by clamping a piece of waste wood on the other side of the board where the drill will come out. The drilling can then be done from one side only with no danger of splintering the wood.

NOTES:

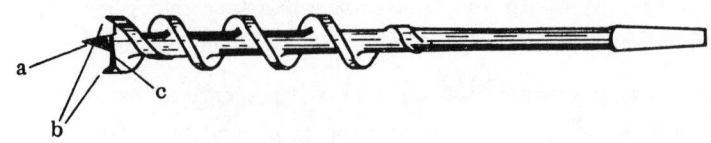

Fig. 1

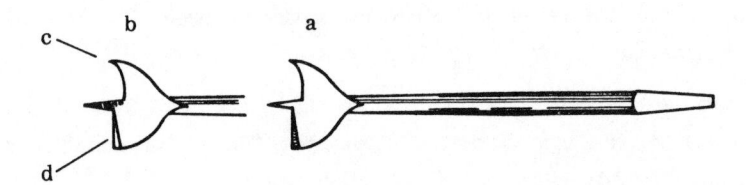

Fig. 2

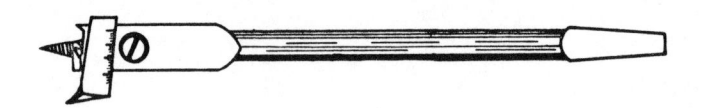

Fig. 3

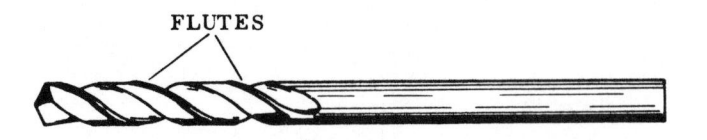

FLUTES

Fig. 4

N P V C		BITS AND DRILLS.
70	TOOLS	

BITS AND DRILLS

Boring bits are used with a brace to bore holes in wood. They have a square-head shank, which fits the chuck of the brace. The shank below the head is cylindrical.

Boring drills usually have a cylindrical shank which fits into a special kind of hand operated drilling machine.

The size of the bit or drill is stamped on the shank, in mm.

AUGER BITS

The twisted part of the auger bit (Fig. 1) guides it and removes the waste. The screw-nose (a) draws the bit into the wood, the two spurs (b) scribe the diameter of the hole, and the cutter (c) cuts it.

Auger bits are used to drill deep holes. They come in sizes from 4 to 40 mm.

CENTRE BITS

The centre bit is used for boring holes in thin timber only, because it doesn't guide itself as the auger bit does.

Centre bits are available with either a point or a screw-nose (Fig. 2a or b). The screw-nose type is preferred over the type with a point, because it draws the bit into the wood.

The spur (c) cuts the rim of the hole and the router (d) removes the waste.

EXPANSION BITS

These bits are used for drilling large holes. The cutter is adjustable with a screw (Fig. 3). The size of the hole can range from 13 to 40 mm in diameter. Special types can expand up to 80 mm and more.

TWIST DRILLS

These drills have twisted flutes (Fig. 4) to bore clean holes in hard or soft woods. One of the main advantages of the twist drill is that it is available in sizes from 1 mm and up.

Besides the drill for wood, there are harder types which can drill holes in metal, stone and concrete. Even though they may look the same, you should never use woodworking drills for metal work.

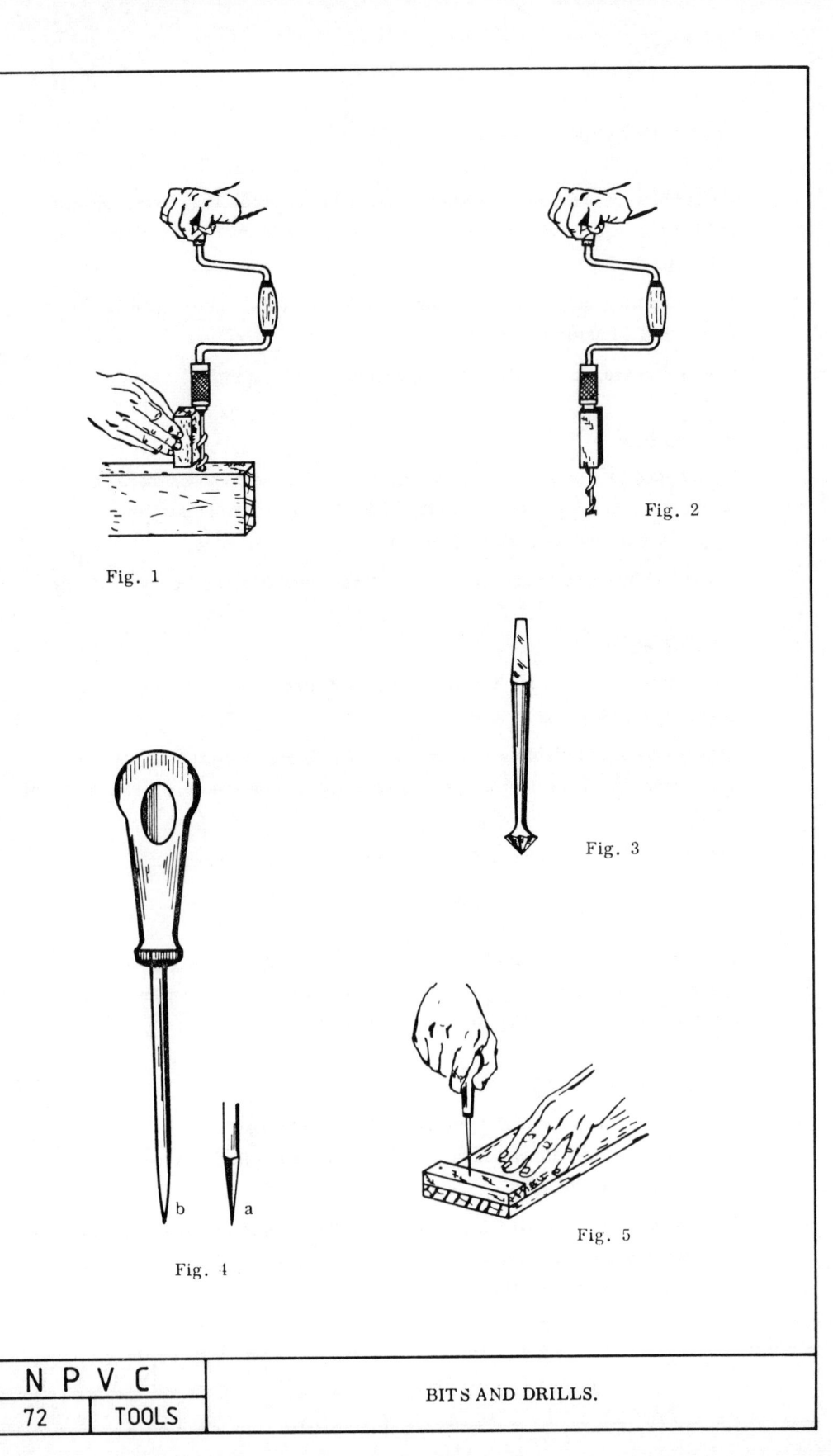

Fig. 1

Fig. 2

Fig. 3

Fig. 4
b   a

Fig. 5

BITS AND DRILLS.

DEPTH STOPS

If you are drilling a number of holes at the same depth, you can save time by cutting a wooden block to the correct size and using it as a gauge (Fig. 1), or boring a hole through the block and fitting it on the bit as shown in Fig. 2.

COUNTERSINKS

Countersinks are used to enlarge the top of a screw hole so that the screw head can fit below the surface of the wood. The most common type is the rosehead countersink (Fig. 3).

AWLS

An awl is a thin, pointed steel rod, which is fitted with a wooden or plastic handle.

Awls are used for marking or piercing holes in wood. The tip can be either square or rounded (Fig. 4a or b). Awls with square shaped tips are preferred for piercing holes for small screws or nails.

- Force the awl into the wood with a turning motion, left and right, so that it cuts its way through the wood (Fig. 5).

- An awl can easily be made from a thin steel rod, by hammering one end to a square shape and sharpening it, then fitting a handle to the other end.

NOTES:

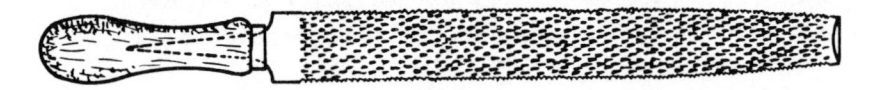

Fig. 1

Fig. 2

Fig. 3

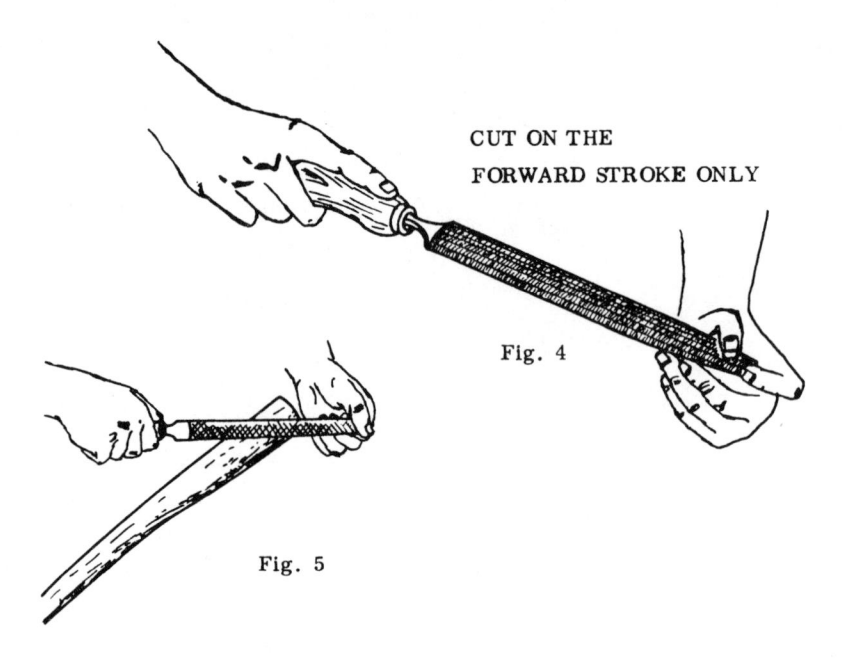

CUT ON THE FORWARD STROKE ONLY

Fig. 4

Fig. 5

N P V C		RASPS AND FILES.
74	TOOLS	

RASPS AND FILES

Rasps (Fig. 1) and files (Fig. 2) are used in woodwork for smoothing wood which cannot be worked easily with any other kind of cutting tool.

- NOTE: Special metalworking files are used to work metal and sharpen tools. Never use woodworking files on metal.

The difference between the rasp and file is in the cut. Files have a series of chisel cuts across their surface, while rasps have many separate small teeth. The rasp is more coarse and cuts faster and rougher than the file.

The handles on both rasps and files must be firmly seated.

Files and rasps are available in different grades of coarseness and in different shapes. Some common shapes for files are (Fig. 3): square, flat, half-round, round and triangular (sawfiles). Rasps are usually half-round or round in section.

HOW TO USE RASPS AND FILES

- Hold the tool at a slight angle to the direction of filing (Fig. 4). By doing this, you file over a wider area and avoid making a hollow spot in one place.

- The actual cutting is done on the forward stroke only.

- When you are filing a curved piece, give the file a sideways sliding motion at the same time as you move it forwards. This is so that the file follows the curve better and doesn't produce flat spots (Fig. 5).

- The rasp is used first, to get the rough shape as quickly as possible. Then the file is used to remove the coarse marks left by the rasp.

- Do not clean rasps or files with sharp tools or steel brushes. To clean them, scrub them with a hard-bristled brush.

NOTES:

Fig. 1

Fig. 2

Fig. 3

Fig. 4

Fig. 5

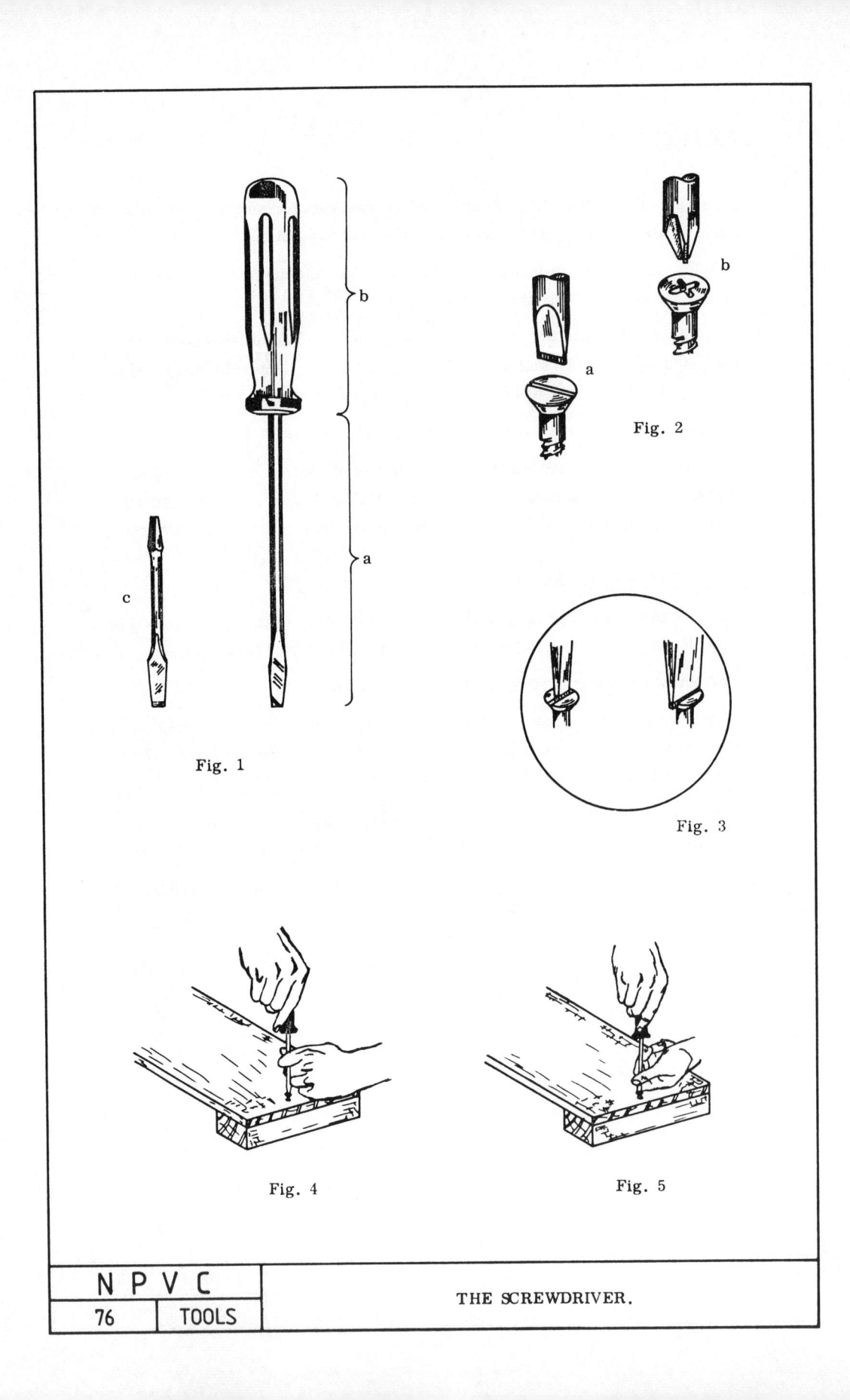

N P V C		THE SCREWDRIVER.
76	TOOLS	

THE SCREWDRIVER

Screwdrivers are used for inserting and removing screws. Many different types of screwdrivers are available. There are also screwdriver bits (Fig. 1, c) to be used with a brace.

The parts of the screwdriver are (Fig. 1): the blade (a) and the handle (b). The blade is either flat or cylindrical and the tip is either ground to a flat straight edge (Fig. 2a) or has a Philips shape (Fig. 2b) to fit different screws. The other end of the blade is shaped to a tang to fit into the handle.

The handle is made out of tough wood or plastic and is sometimes fitted with a ferrule to prevent splintering and keep the blade from turning in the handle.

The size of the screwdriver is determined by the length of the blade from the tip to the handle and by the diameter of the blade.

In order to work efficiently and not damage the slots of the screwheads, select a screwdriver with a tip that is the same width as the slot of the screw. A tip which is too small can slip and damage the slot, while one which is wider than the screwhead can scrape and damage the workface around the screw (Fig. 3).

The screwdriver tip must be properly formed (see Maintenance of Tools). A badly formed tip will cause the screwdriver to slip out of the slot and damage it.

HOW TO USE THE SCREWDRIVER

- Grasp the handle firmly in your right hand with your palm resting on the end of the handle; the thumb and forefinger extend along the handle.
- While the right hand changes grips to turn the handle, the left hand steadies the tool and keeps it in the slot. Figs. 4 and 5 show two methods of using a screwdriver.

NOTES:

Fig. 1

Fig. 2

Fig. 3

a

Fig. 4 b

45° 45°

Fig. 5

45° 90° 45°

Fig. 6

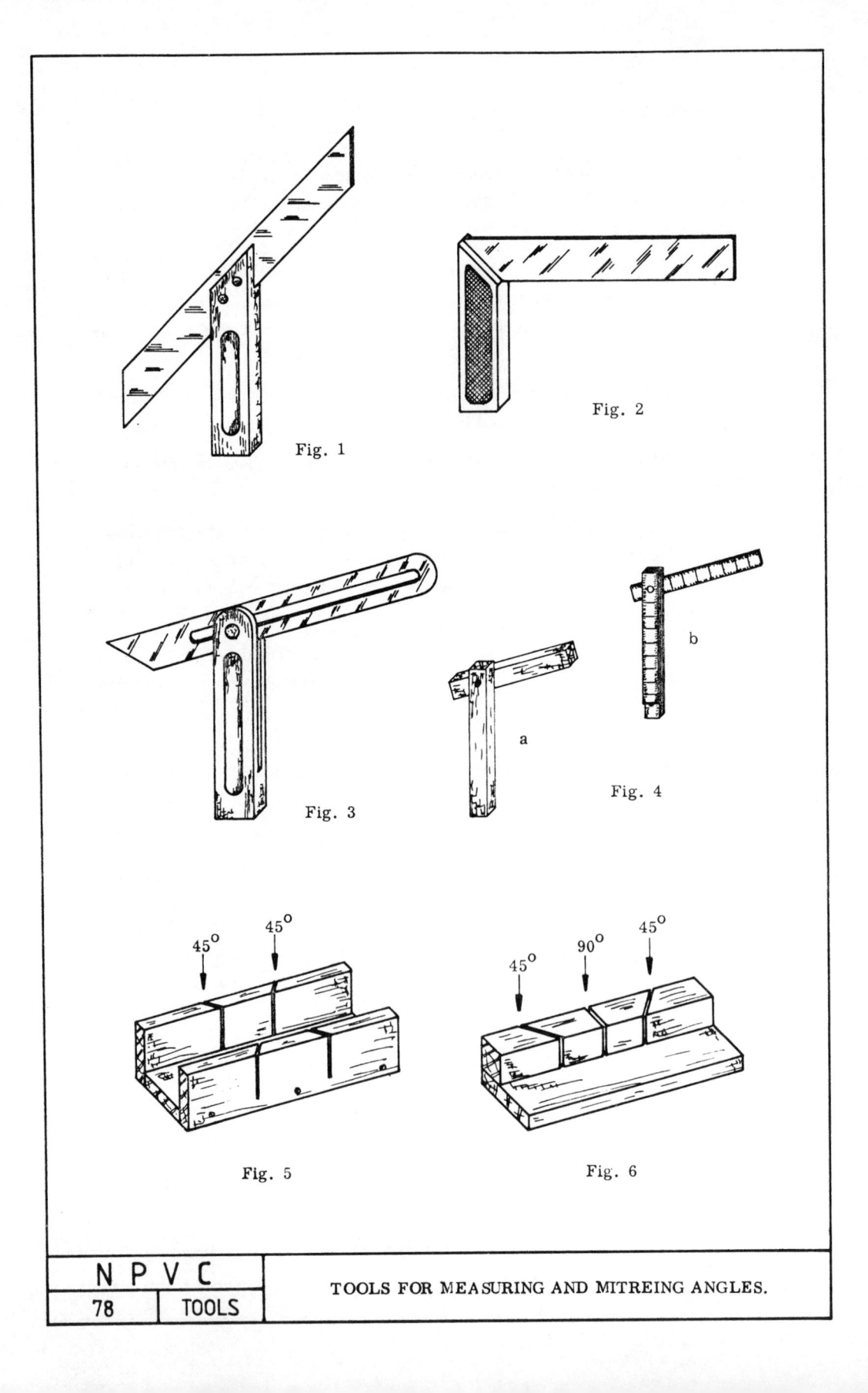

NPVC	
78	TOOLS

TOOLS FOR MEASURING AND MITREING ANGLES.

TOOLS FOR MEASURING AND MITRING ANGLES

THE MITRE SQUARE

This is used to mark and test angles of 45 and 135 degrees. The blade is fixed at 45 degrees to the stock (Fig. 1).

THE TRY AND MITRE SQUARE

This is a combination of a try square and a mitre square. The end of the stock where it meets the blade is cut at 45 degrees (Fig. 2), so the square can be used for setting out and testing angles of 45 and 135 degrees, as well as 90 degrees.

THE SLIDING BEVEL

This is an adjustable square for marking out, testing and duplicating angles from 0 to 180 degrees. It has a stock and a slotted blade which can be adjusted to any angle and is held in place by a screw or a wing nut (Fig. 3).

A simple sliding bevel can be made by fixing two pieces of wood together with a nail or screw (Fig. 4a). Another method is to use the first section of a folding rule (Fig. 4b).

THE MITRE BOX

The mitre box (Fig. 5) is built of three pieces of wood, one forming the base and two parallel sides. It has saw kerfs in the sides at 45 degrees to the left and right to guide the saw in cutting mitres (cutting at a 45 degree angle).

THE MITRE BLOCK

This is used to mitre small sections of wood accurately. It is made of two pieces of wood with three cuts in the top piece; 45 degree cuts left and right and a 90 degree cut in the centre to help in sawing accurately square (Fig. 6).

- Mitre boxes and mitre blocks should be made of very hard wood and the saw cuts should be made with the same saw which will be used in them.

NOTES:

Fig. 1

Fig. 2

Fig. 3

Fig. 4

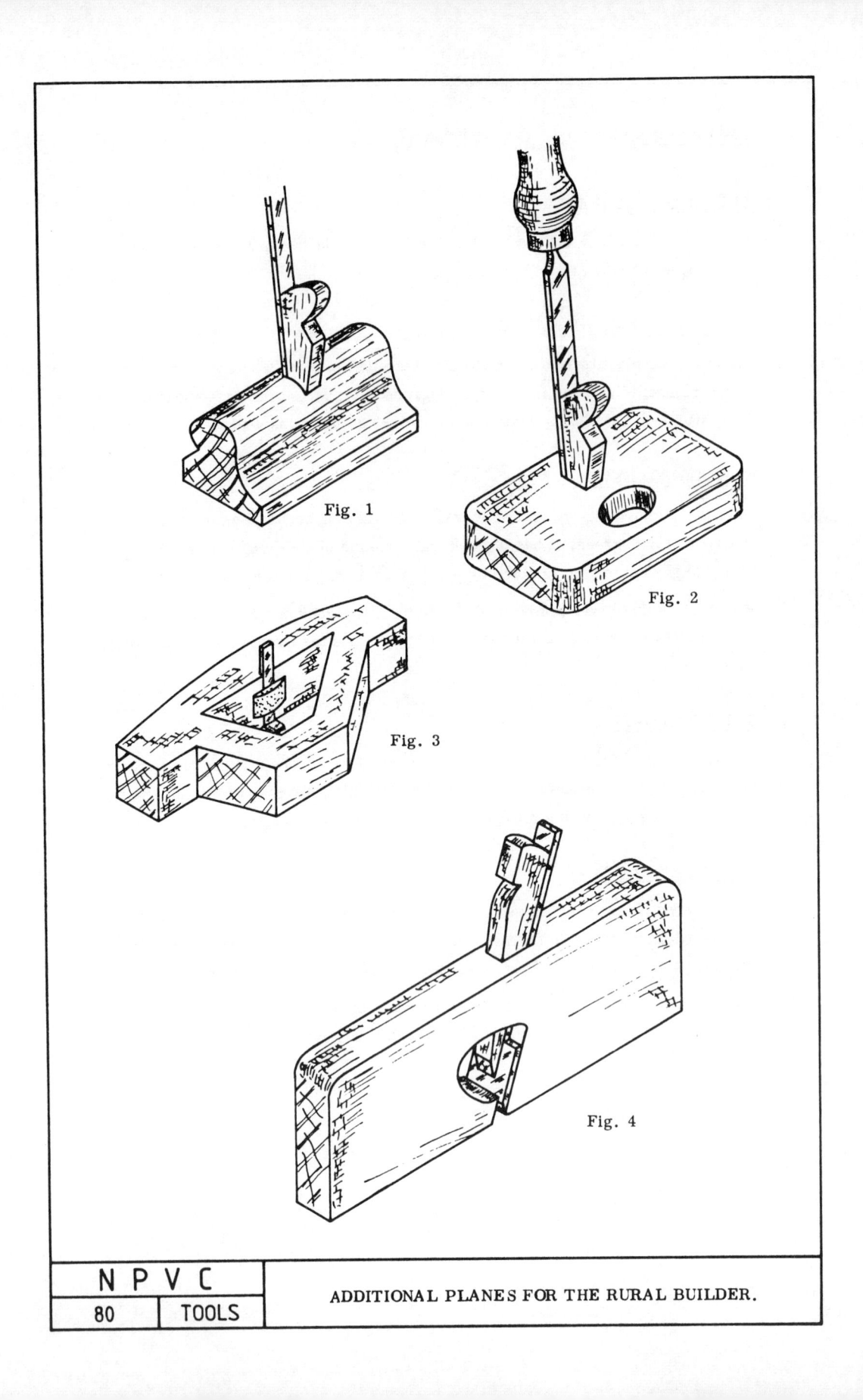

ADDITIONAL PLANES FOR THE RURAL BUILDER.

ADDITIONAL PLANES FOR THE RURAL BUILDER

THE ROUTER PLANE

This is a tool adapted from the "old woman's tooth" (Fig. 1) shown on the left page. It is used for planing the bottoms of grooves, trenches, etc. after they have been chopped out with a chisel and mallet.

There is a wooden type as well as a metal type. The wooden type has the advantage that it can be made locally (Fig. 3).

For our Rural Building course, the old woman's tooth has the advantage that an ordinary chisel can be used as a cutting iron (Fig. 2), making it unnecessary to obtain a special iron.

THE REBATE PLANE

The rebate plane is used for working along the edge of timber. It is used for making rebates as well as for cleaning them up.

The use of the rebate plane is described in the section on Rebated Butt Joints, in Rural Building, Basic Knowledge, page 132).

NOTES:

Fig. 1

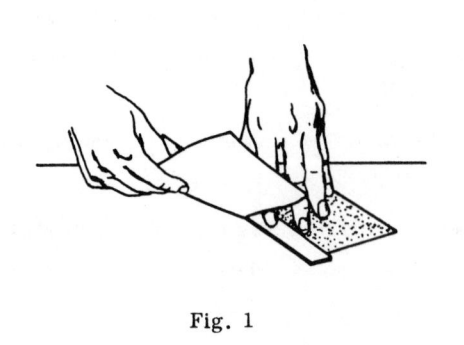

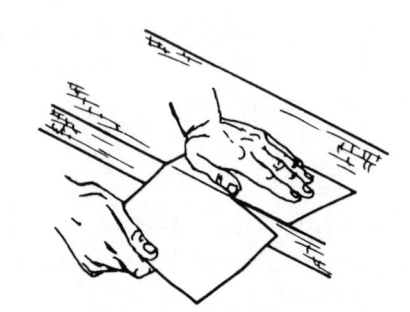

Fig. 2

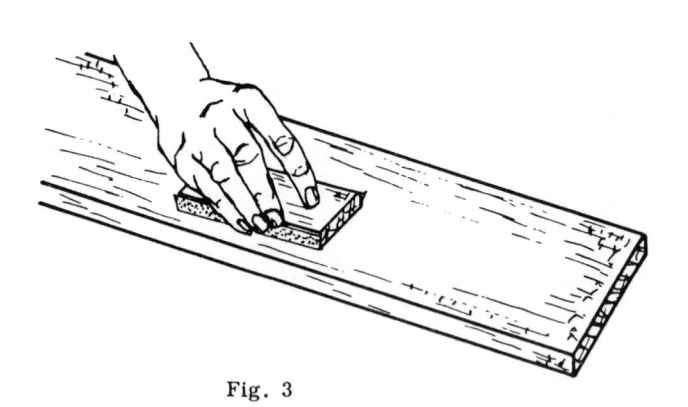

Fig. 3

N P V C		SANDPAPER.
82	TOOLS	

SANDPAPER

Sandpaper is used to smooth wood surfaces, remove old paint, etc. It is used after all the planing work is done; sandpaper cannot be used instead of the smoothing plane.

Sandpaper is made from grains of very hard material glued to paper. The sharp edges of the grains are what cut into the surface of the wood. The sharper the grains, the better the sanding effect.

Sandpaper is graded according to the space between the grains: the more widely spaced the coarser the grade. Use coarse grades for rough surfaces or for the first sanding and finer grades for the final sanding. The commonly used grades are from No. 00 (fine) to No. 2 (coarse). Usually No. 1/2 or No. 1 is satisfactory for coarse sanding on wood and No. 0 for the final or finished sanding.

HOW TO USE SANDPAPER

- Tear the paper to the right size by creasing it and then tearing it over the edge of a bench or ruler (Figs. 1 & 2).

- Wrap it part way around a flat block of wood (Fig. 3).

- For ordinary sanding, move the block back and forth with the grain; not in a circular motion or across the grain, which will roughen and scratch the wood instead of smoothing it (Fig. 3).

- If you are sanding off old paint, this may be done across the grain.

- Keep the block flat to the wood surface, particularly on narrow edges and be careful not to round the corners.

- Use only moderate pressure on the sandpaper block. Too much pressure may cause the paper to wrinkle or tear. Keep the sandpaper free of dust by knocking and shaking it out often.

NOTES:

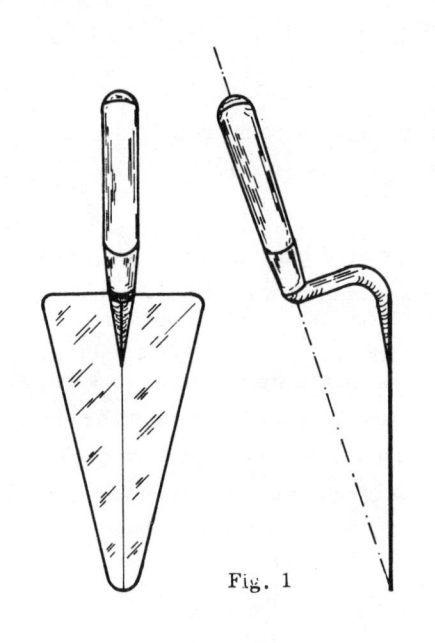

Fig. 1 **TRIANGULAR TROWEL**	Fig. 2 — a **RECTANGULAR TROWEL**
Fig. 3 **HEAVY TROWEL**	Fig. 4 **STEEL FLOAT**

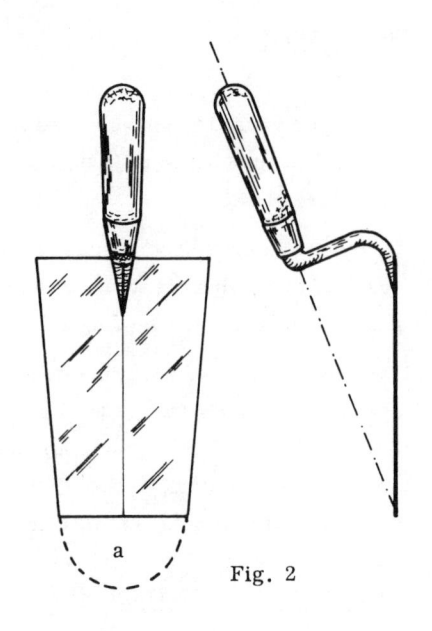

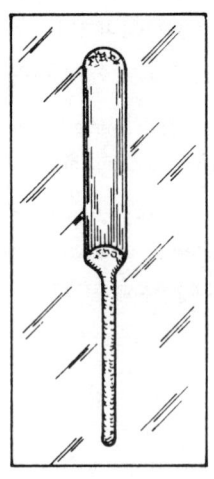

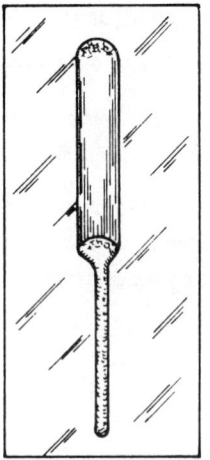

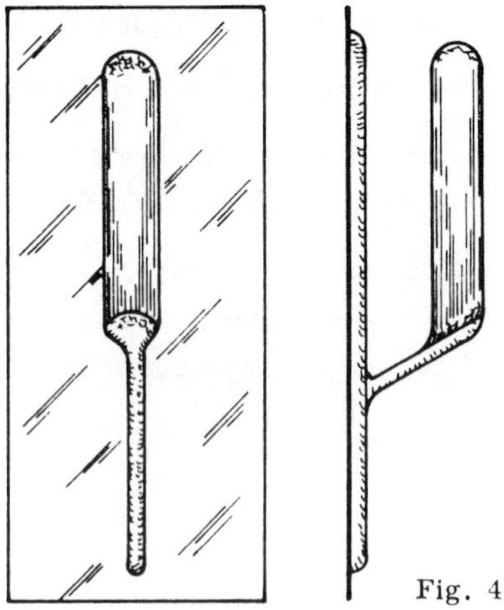

NPVC	
84	TOOLS

OPTIONAL TOOLS.

OPTIONAL TOOLS

There are many useful tools which the apprentice will eventually come into contact with and which perhaps he would like to purchase.

Although the set of tools supplied to the Rural Builder will be adequate to do the work, a good craftsman will always try to improve upon his tool set.

Therefore, the purpose of the following introduction to some additional tools is simply to round out the apprentice's general knowledge about tools.

OTHER TYPES OF TROWELS

Fig. 1 shows a triangular trowel, while the one in Fig. 2 is almost rectangular. The rectangular type of trowel is also available with a rounded tip (Fig. 2a). All three types of trowel are useful in Rural Building and the decision of which one to use depends merely upon their availability and the Builder's personal preference.

Fig. 3 shows a heavier trowel designed specially for concrete work. The head is rounded to make it easier to pick up concrete. The blade is rather thick compared to other trowels and the straight shank is connected to the blade at a 90 degree angle to reduce the flexibility of the trowel.

STEEL FLOAT

The steel float consists of a thin rectangular blade about 12 cm wide by 28 cm long. There is a handle fitted to the back side of the blade (Fig. 4).

This is a finishing tool used for smoothing surfaces such as floors and plaster.

NOTES:

PLOUGH PLANE

Fig. 1

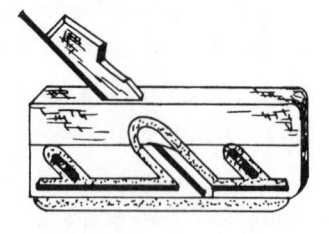

FILLISTER PLANE

Fig. 2

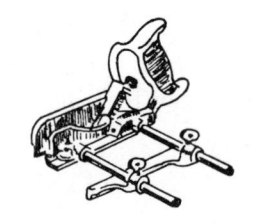

COMBINATION PLANE

Fig. 3

METAL PLANE

Fig. 4

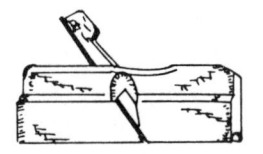

MOULDING PLANE

Fig. 5

SPOKE SHAVE

Fig. 6

OPTIONAL TOOLS.

OTHER PLANES

There are many varieties of planes available for different jobs. However, in Rural Building, we can do a good job with the four different planes we have and we will leave it to the future craftsman to decide if it is necessary to obtain any other plane in addition.

Below we describe briefly some other planes which might be a choice for the future builder. They are shown on the left page.

The plough plane (Fig. 1) is used to make grooves in wood.

The fillister plane (Fig. 2) is used for making rebates.

The combination plane (Fig. 3) is a combination of the fillister plane, plough plane and moulding plane.

There are various types of metal planes (Fig. 4) used for different purposes in furniture making.

The moulding plane (Fig. 5) resembles a rebate plane and it is used for making profiles.

The spoke shave (Fig. 6) is used to true and smooth edges after sawing.

NOTES:

NOTES:

PART 2: MAINTENANCE OF TOOLS

The importance of good tool maintenance is something which is readily appreciated by everyone; the beginner will quickly realize that without maintenance, the finest tools are no more useful than the most inferior ones.

Apart from their general maintenance in terms of cleanliness, rust prevention and avoidance of damage from rough handling, the most important aspect of maintaining tools is in the preparation and preservation of good cutting edges.

Until one has some experience in tool maintenance, one easily overlooks the fact that from the moment that a fine cutting edge is formed, it becomes the most delicate and easily damaged part of the tool. Even the lightest touch of another piece of metal can spoil the edge, wasting the work which has gone into making it.

Such damage can be avoided in commonsense ways such as:
- By keeping the bench clear of tools which are not in use.

- By laying planes down on their sides or keeping the toe raised on a wood support so that the plane iron does not touch the bench top.

- By keeping chisels and saws in the box when they are not in use.

- By keeping the bench clear of ironmongery such as nails, screws, hinges, etc.

- And by putting tools down on wooden surfaces only.

NOTES:

CUTTING ANGLE

Fig. 1

70°

30°

2°

Fig. 2

Fig. 3

Fig. 4

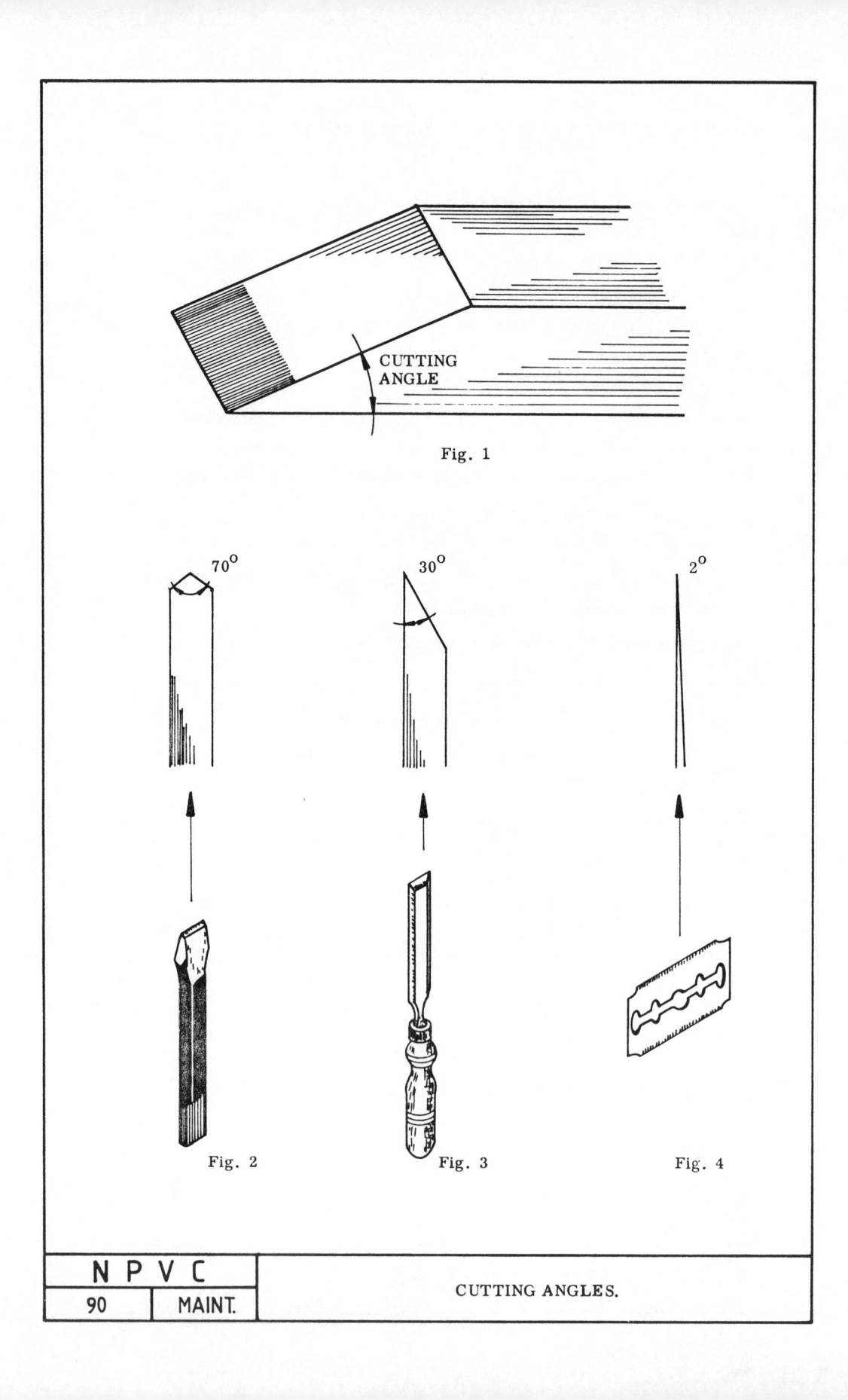

N P V C		CUTTING ANGLES.
90	MAINT.	

CUTTING ANGLES

Before you can properly maintain your tools, you need to know something about the shape of the cutting edges on tools like plane irons and chisels. A cutting edge is formed where the two faces of a wedge come together at an angle, which is the cutting angle (Fig. 1).

In general, the harder the material that you want to cut, the larger the cutting angle you need. For example, to cut steel or stone you need a tool with a cutting angle of about 70 degrees, like the cold chisel (Fig. 2). To cut a soft material like leather, a knife or a razor blade with a cutting angle of only a few degrees (Fig. 4) will work efficiently. Of course, you would not be able to cut steel with a razor blade.

In both these cases, it is important that the cutting edge of the tools is sharp, although they have different cutting angles.

For cutting wood, we generally have one standard cutting angle for the tools like plane irons and chisels, although different kinds of wood can vary quite a bit in hardness. This standard angle of between 25 and 35 degrees is more or less suitable for all types of wood (Fig. 3). The craftsman will find with experience the right cutting angle for his needs.

Note that the smaller the cutting angle, the more easily the tool will cut and also the sooner it will become dull.

NOTES:

Fig. 1

x
2 x

ENLARGED DETAIL

WRONG   RIGHT

Fig. 2

Fig. 3

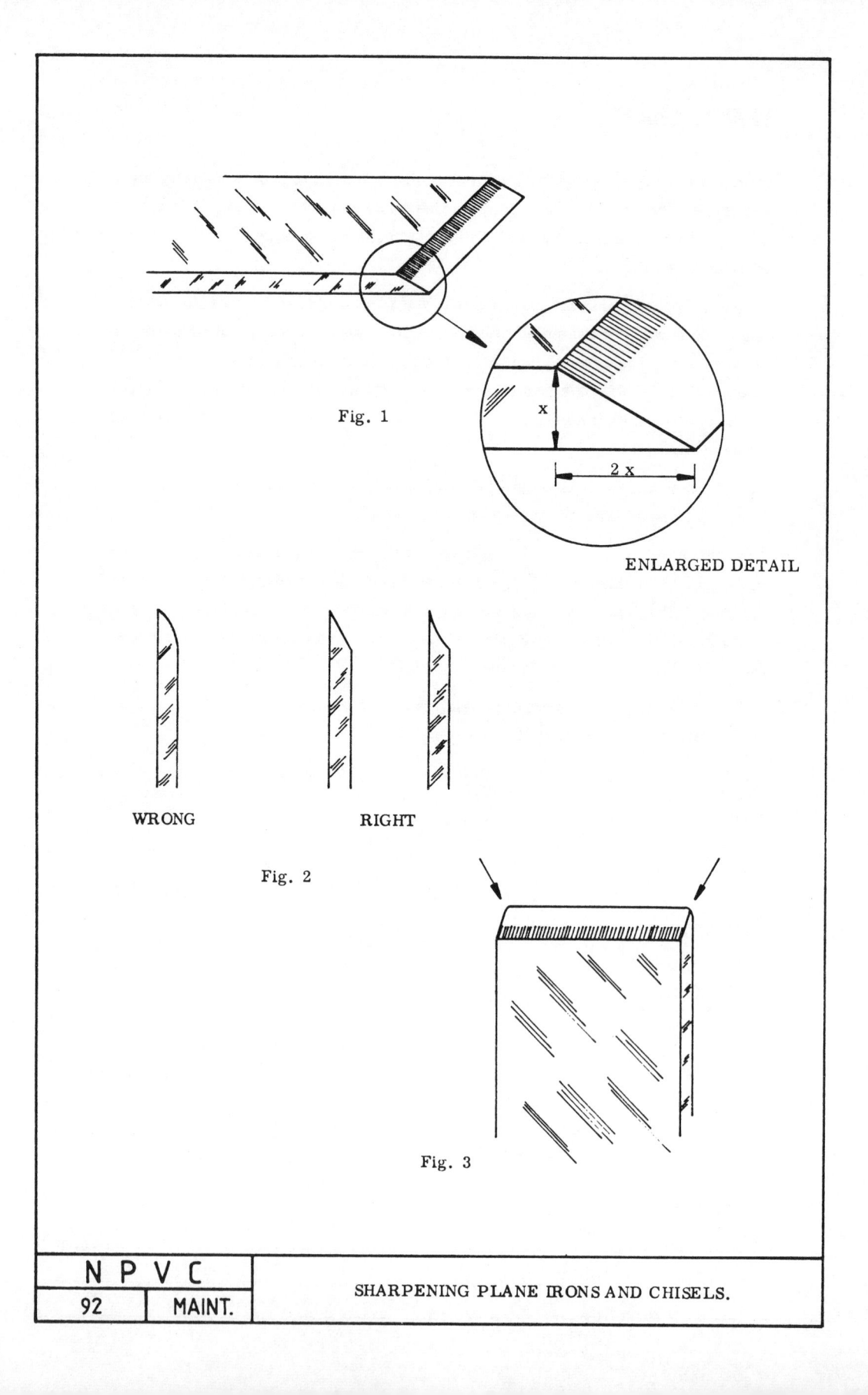

NPVC		SHARPENING PLANE IRONS AND CHISELS.
92	MAINT.	

SHARPENING PLANE IRONS AND CHISELS

Two different operations may be needed to produce a keen edge on these tools: grinding and honing.

Simply honing tools on a sharpening stone is all that is needed to sharpen them if they are dull but undamaged.

If the edges have been damaged, or worn down by many honings, they will need to be ground first and then honed. The grinding step is done on the rough side of the sharpening stone, or if the damage is very bad, on the grinding wheel.

ANGLE AND SHAPE OF THE CUTTING EDGE

- For normal work, grind plane irons and chisels so that the length of the bevel is a little more than twice the thickness of the blade (Fig. 1). This gives an angle of 25 to 30 degrees.

- The bevel should be ground straight or slightly concave, not rounded or convex (Fig. 2).

- For jack planes it is desirable to grind the corners of the blade slightly rounded as shown in Fig. 3. This lets the iron cut thicker shavings without causing deep grooves or plane marks in the wood surface.

- The cutting irons of smoothing planes, rebate planes and chisels should not be rounded.

- When you sharpen the cutting iron of a plane, always remove the cap iron first.

NOTES:

Fig. 1

Fig. 2

Fig. 3

Fig. 4

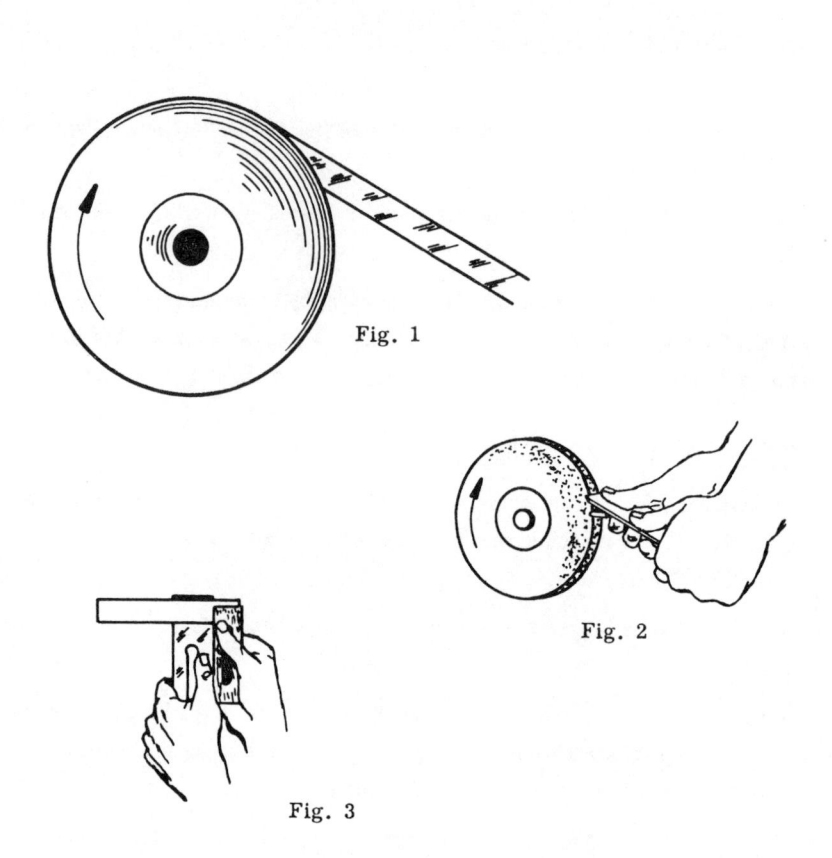

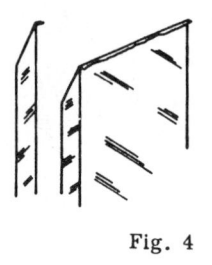

Fig. 5

NPVC 94 MAINT.   SHARPENING PLANE IRONS AND CHISELS.

HOW TO GRIND PLANE IRONS AND CHISELS ON THE GRINDING WHEEL

When the cutting edge of a tool is badly damaged, it is ground on the grinding wheel. This is done by holding the tool at a constant angle to the rotating edge of the wheel (Figs. 1 & 2).

Each grinding makes the tool a bit shorter, and thus shortens its life. Use your tools carefully so that they don't need to be reground as often.

When you grind a plane iron or chisel on the grinding wheel, the following points are important to keep in mind:

- Before grinding, test the cutting edge for squareness. Put a try square on top of the chisel or plane iron, with the cutting edge projecting slightly past the try square blade (Fig. 3). Be careful not to touch the cutting edge with the square.

- Hold the tool against the wheel in a manner that will produce a smooth, even bevel, with the desired angle.

- If possible, adjust the work rest of the grinding wheel, so that when the tool is held firmly against the rest it will come into contact with the wheel at the correct angle (Fig. 2).

- Grasp the tool so that your first finger is against the work rest; this will enable you to replace the tool in the proper position after removing it for inspection or to dip it in water (Fig. 5).

- During grinding, a wire edge or burr is formed on the tool. You can feel this burr if you run your thumb across the edge at the back (Fig. 4).

- It is important that the grinding wheel always turns towards you (Fig. 1), so that the burr that is formed remains on the blade. If the wheel were rotating away from you the burr would tear off, leaving an uneven edge.

- Turn at a moderately fast speed, not so fast that the gears whine or the grinder vibrates.

- Always work with another person on the grinding wheel, so that he can turn the wheel while you do the grinding.

- Hold the tool against the wheel with a medium firm pressure.

- Move the tool from side to side across the face of the wheel.

- Take care that the cutting edge does not become overheated and thus softened during grinding. Prevent overheating by frequently dipping the cutting edge into a tin of water.

SHARPENING PLANE IRONS AND CHISELS.

- Inspect the edge often, to see if the tool is being ground to the proper shape and angle.
- With a bit of practice, you can check the cutting angle by eye, remembering that the length of the bevel should be a little more than twice the thickness of the blade. Use a rule at first to check the length of the bevel and the thickness of the blade.
- Continue grinding until the dull edge is removed, all the marks are removed, the edge is straight and square and the bevel has the required angle. If the edge is not square, correct it during the grinding by pressing carefully more on one side than on the other.
- Remove the burr or wire edge left by the grinding wheel by honing the tool on a sharpening stone.

NOTES:

GRINDING ON THE COARSE SIDE OF THE SHARPENING STONE

If the tool has a cutting edge with only a few marks in it and no serious damage, it can be ground on the coarse side of the sharpening stone, instead of using the grinding wheel.

- Soak the stone in water and grind the bevel of the tool on the coarse side of the stone, until a slight burr or wire edge is formed.

- When grinding, place the cutting iron on the stone so that the bevel lies flat (Fig. 1) and rub it with circular movements. Do not rub it just back and forth, as this makes the stone wear unevenly.

- Be particularly careful to move your hands parallel to the surface of the stone and do not allow them to make a dipping movement, as this will round the cutting edge.

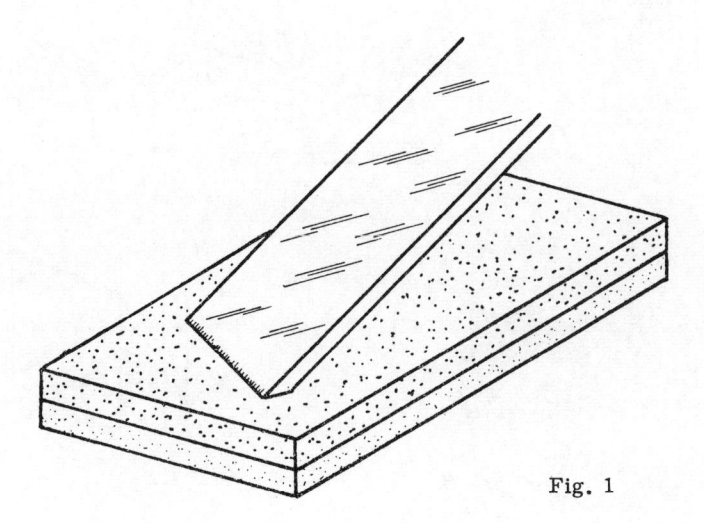

Fig. 1

NOTES:

SHARPENING PLANE IRONS AND CHISELS.

Fig. 1

Fig. 2

Fig. 3

Fig. 4

Fig. 5

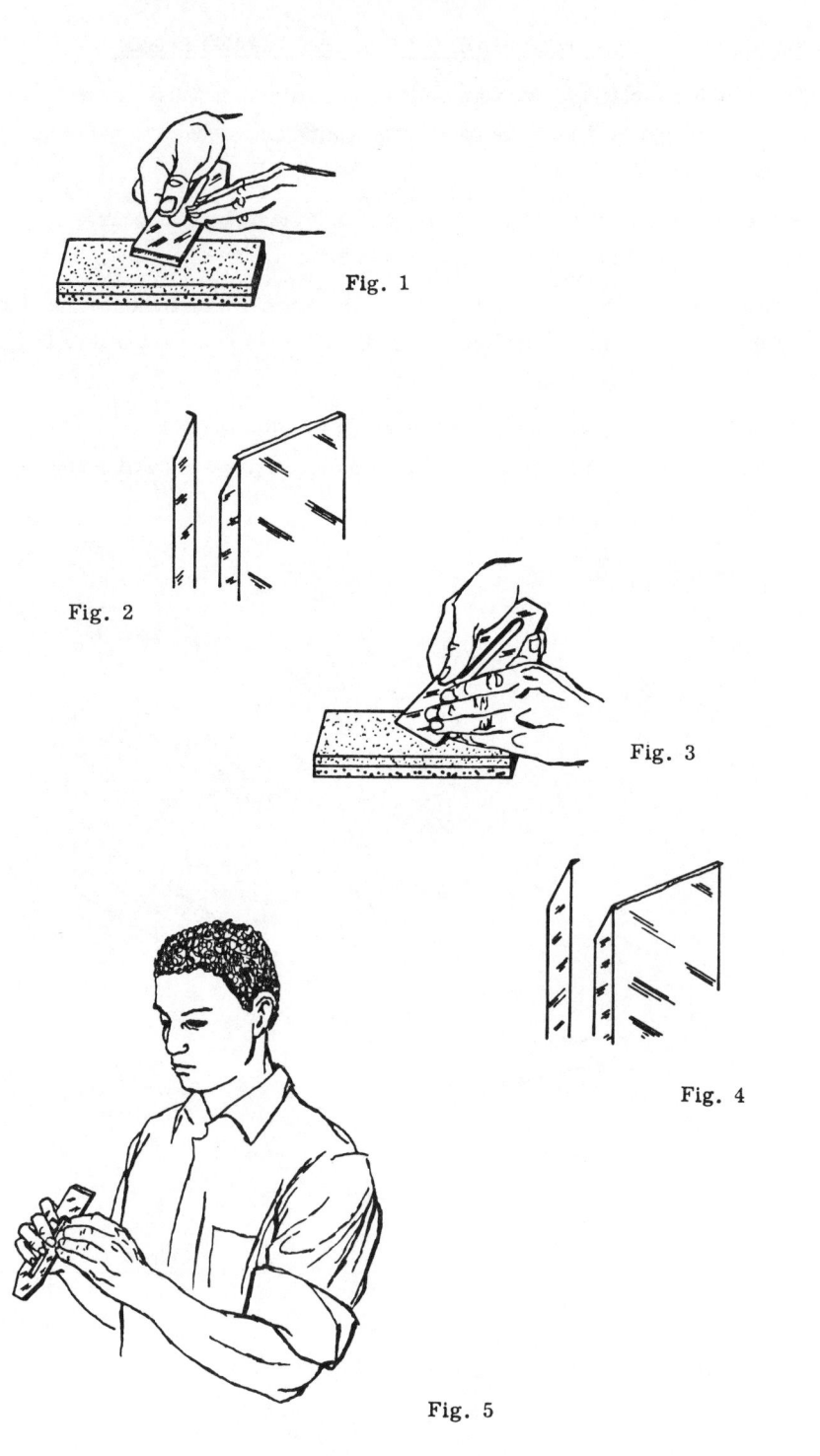

N P V C	
98	MAINT.

SHARPENING PLANE IRONS AND CHISELS.

HOW TO HONE PLANE IRONS AND CHISELS

Honing is done on the fine side of the sharpening stone. Honing produces a keen edge.

After a slight wire edge has been produced on the coarse side of the sharpening stone, or on the grinding wheel, it is removed by honing on the fine side of the stone.

- First, soak the stone in water. Place the tool perfectly flat on the stone with the bevel up and push it forward (Fig. 1). A few strokes will turn the burr from the flat side of the tool to the side with the bevel (Fig. 2).

- When the wire edge turns, turn the tool over so the bevel is flat on the stone and hone lightly on the bevel (Fig. 3).

- Then reverse the tool again and hone on the flat side.

If the honing is properly done, the wire edge will quickly become smaller and smaller (Fig. 4) and eventually disappear. The tool will then be sharp.

To check whether the tool is sharp, draw the nail of your thumb across the edge of the tool.

Marks, such as may have been caused by a nail, can be detected by holding the iron to the light. A sharp edge cannot be seen, while a dull one will show up in the light and appears as a narrow, shiny surface (Fig. 5).

If the tool is not held perfectly flat when the flat side is honed, a small bevel may be produced on the flat side and it will then be impossible to put the edge in good condition without regrinding it.

- In alternately honing the flat and bevelled sides, make sure that the wire edge is actually turned from the flat to the bevelled side before you reverse the tool for honing on the bevelled side (Figs. 2 & 4).

- To hone a slightly dull edge without grinding it, rub it on the sharpening stone in the same way, but only on the bevel side, not on the flat side.

Chisels are sharpened in the same way as cutting irons. Keep in mind that they are narrow and they should not be worked all the time in the centre of the stone, as this will quickly cause the centre of the stone to become hollow.

NOTES:

Fig. 1

Fig. 2

Fig. 3

Fig. 4

Fig. 5

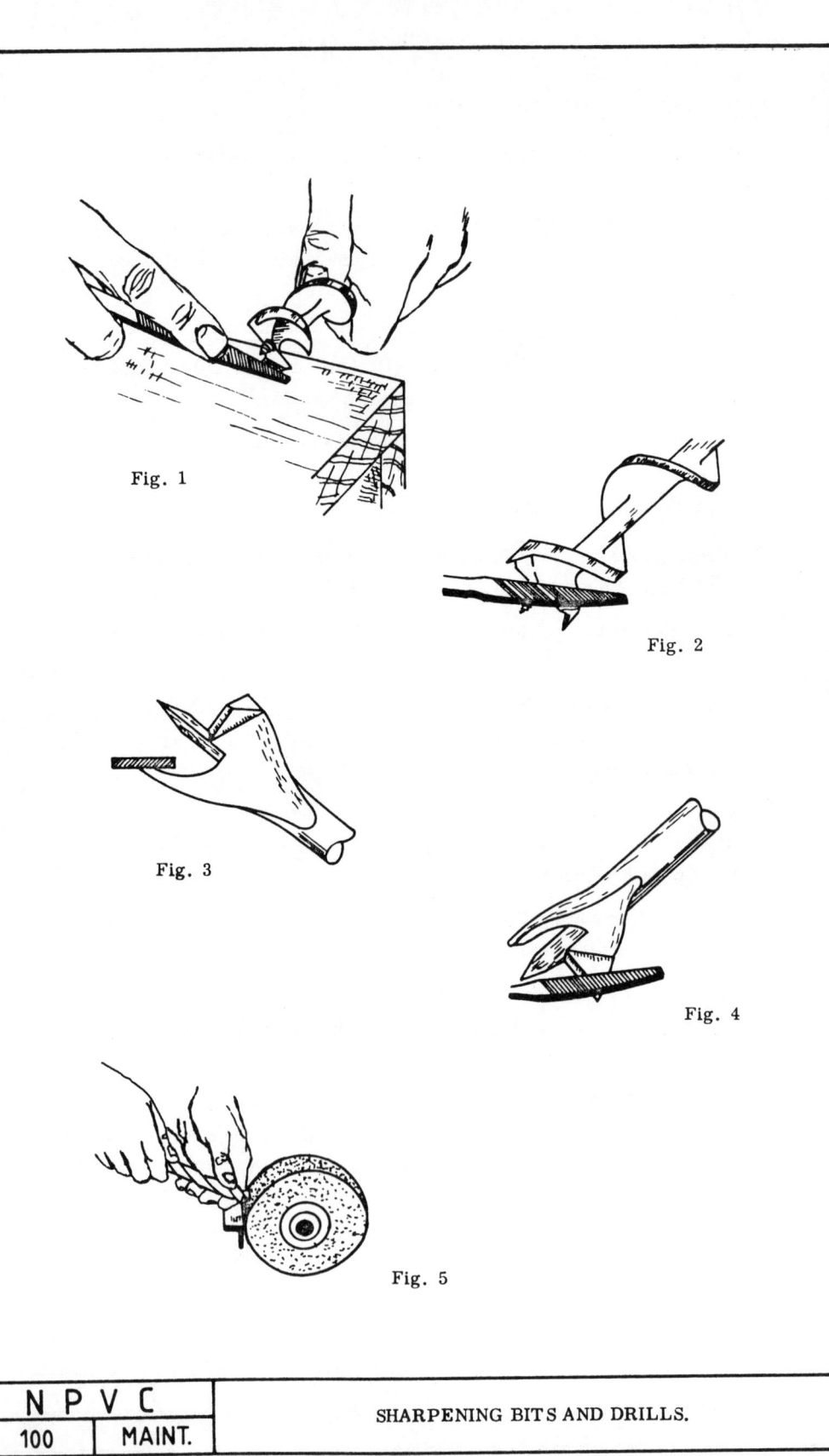

N P V C	
100	MAINT.

SHARPENING BITS AND DRILLS.

SHARPENING BITS AND DRILLS

Manual wood bits are usually sharpened with a small file. It is best to avoid too much sharpening. Your tools will last longer if you use them carefully and keep them in a good case, so they won't need sharpening so often.

SHARPENING THE AUGER BIT

Auger bits are sharpened as shown on the left (Figs. 1 & 2).

- For the spurs, use a small file and sharpen from the inside face only (Fig. 1). The spurs are never sharpened on the outer faces, as this would change their diameter.
- For the cutters, file from the top side (Fig. 2). Retain the original bevel and remove about the same amount of material from each side.
- Hold the bit firmly to the edge of the bench during filing.

HOW TO SHARPEN THE CENTRE BIT

- To sharpen the spur, file from the inside (Fig. 3). To sharpen the router, file the bevel from the top side (Fig. 4).
- When sharpening the centre point, take care that the point remains exactly in the middle between the spur and the outside of the router.

HOW TO SHARPEN THE TWIST DRILL

- Grind the nose from the bottom (Fig. 5). Keep the original shape and angles and remove the same amount from each side.
- Never file the flutes as this would change the diameter of the drill. If the drill is really out of shape, contact someone with metalworking experience.

HOW TO SHARPEN AWLS

- To sharpen an awl you can use the sharpening stone, or if it is damaged rub it down with a file. Pointed awls can be sharpened on a grinding wheel.

NOTES:

Fig. 1

Fig. 2

Fig. 3

Fig. 4

Fig. 5

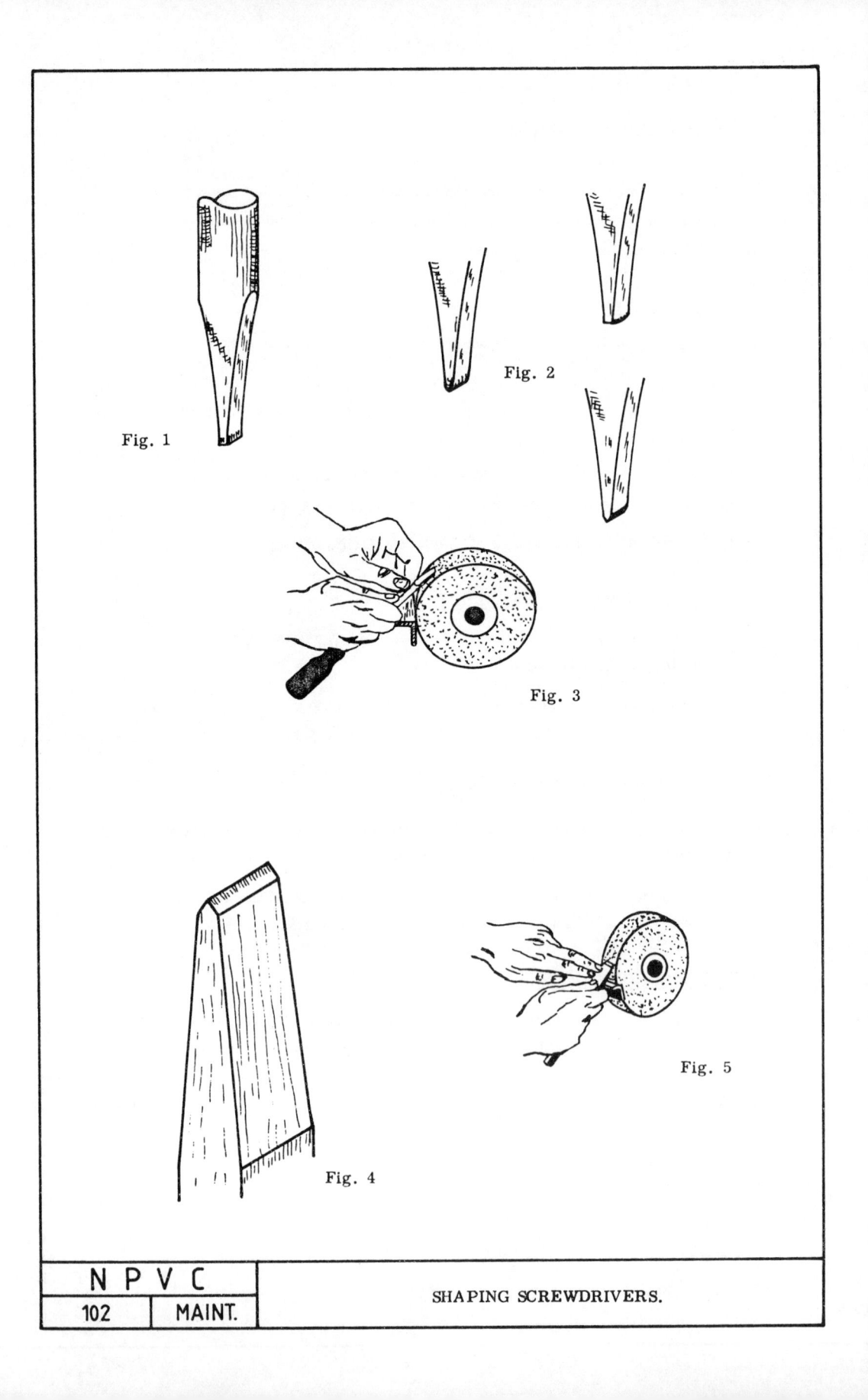

NPVC		SHAPING SCREWDRIVERS.
102	MAINT.	

SHAPING SCREWDRIVERS

A screwdriver should be ground or filed to a very blunt end (Fig. 3). The two flat surfaces should be straight and parallel near the tip. The end should be square to the flat sides, and should be a little less thick than the width of a screw slot (Fig. 1).

If the end is rounded or sharpened to a knife edge, it will easily slip out of the screw slot and damage the slot (see Screwdrivers, page 77). In Fig. 2 there are some examples of badly shaped screwdriver tips.

SHAPING COLD CHISELS

Cold chisels should be ground or filed with the bevels on the cutting edge making an angle of about 70 degrees to each other (Figs. 4 & 5).

SHAPING BLOCK SCUTCHES

A block scutch should be shaped and filed to an angle of 70 degrees. Unlike the cold chisel, it is filed only from the inside (Fig. 6).

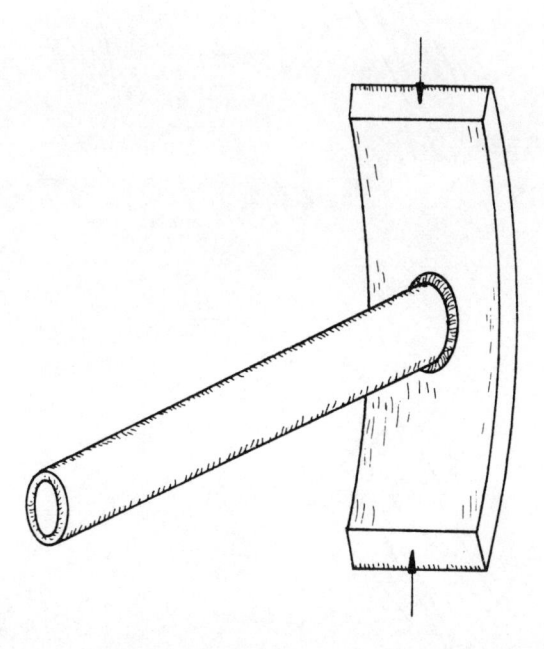

Fig. 6

SHAPING SCREWDRIVERS.

N P V C
MAINT. 103

Fig. 1

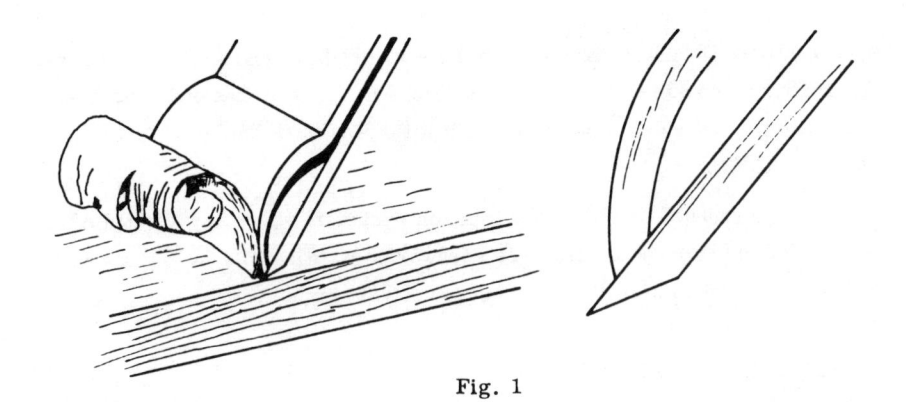

Fig. 2

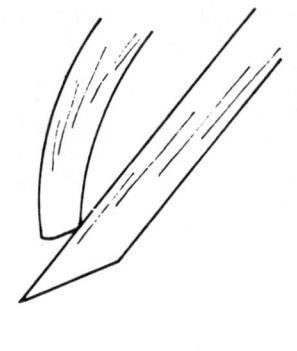

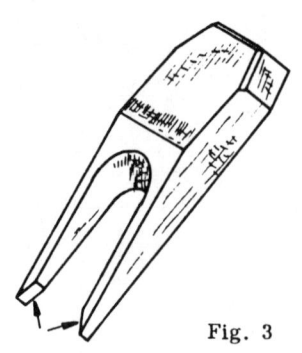

Fig. 3

NPVC
104 MAINT.

MAINTAINING WOODEN PLANES.

MAINTAINING WOODEN PLANES

HOW TO REFACE THE SOLE

Whatever the quality of the wood stock, the soles of all wooden planes eventually wear and require refacing.

Refacing can be done on a large sheet of sandpaper, which is fastened to a true flat surface. The cutting iron of the worn plane is pulled inside the mouth, not removed, for it is best to have the stock held in the usual pressure by the wedge.

- Sand the surface down, test it with a straight edge and winding strips and oil the wood lightly.

ABOUT THE BEDDING OF THE CAP IRON

It is essential that the edge of the cap iron should bed perfectly on the face of the cutting iron when they are screwed together (Fig. 1). Even the slightest gap between the two will allow a shaving to enter and block the mouth of the plane (Fig. 2).

Usually this problem is caused by a fault in the cap iron. The cap iron should be rubbed on a sharpening stone or filed to restore the true edge. Keep the cap iron clean and shiny.

FITTING THE WEDGE

After much use, the wedge may no longer fit well into the stock. Heavy use of the hammer when knocking it in may cause the wedge to become misshapen because of the pressure.

When it becomes difficult to remove the wedge in the usual way (by hitting sharply on the striking button), the wedge should be removed, cleaned and filed or sandpapered to the correct fit again.

The ends of the wedge can become worn and cause the plane to stuff because they are too far inside the plane. If this happens, recut the ends of the wedge (Fig. 3).

NOTES:

Fig. 1

Fig. 2

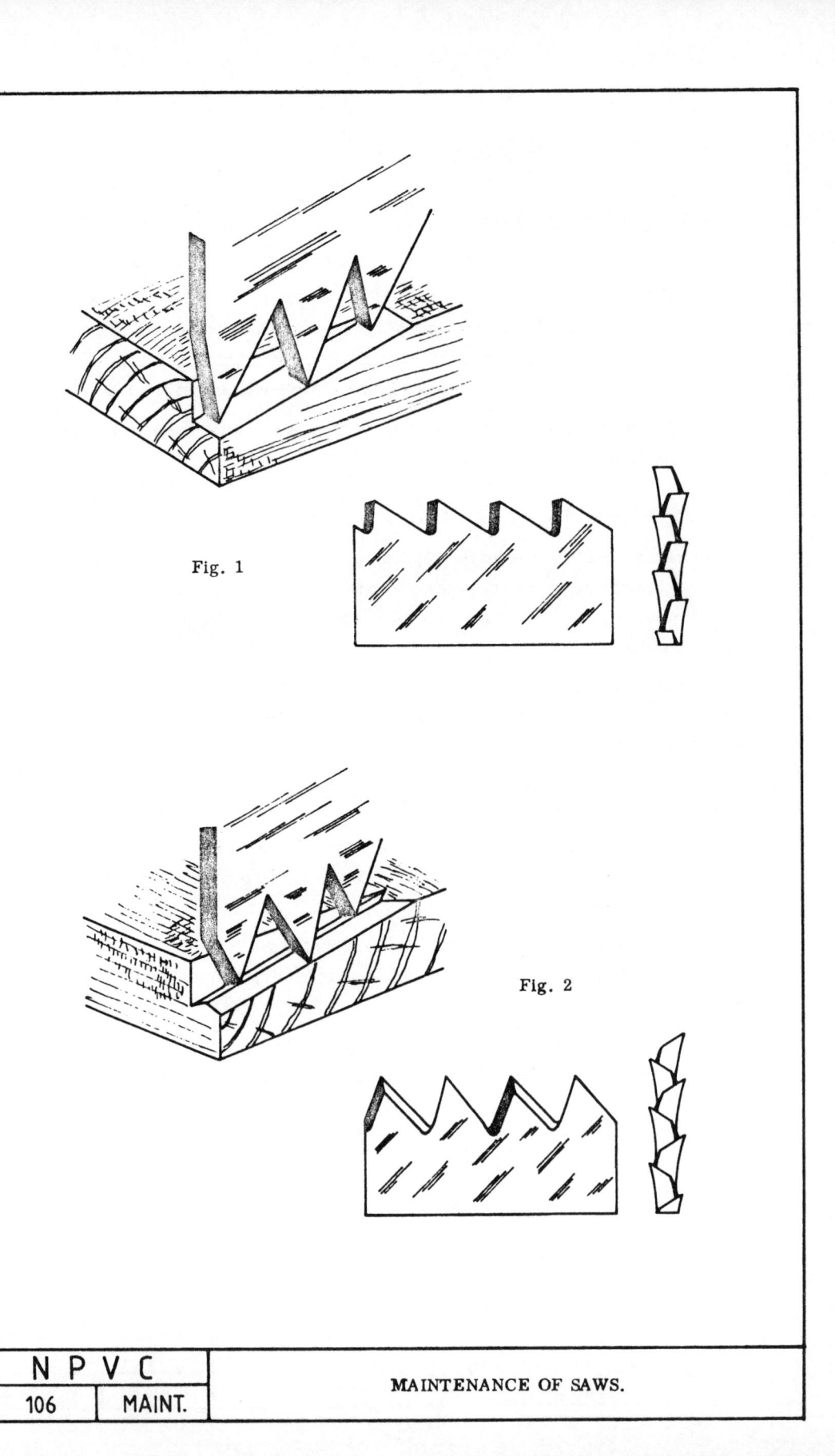

NPVC 106 MAINT.

MAINTENANCE OF SAWS.

MAINTENANCE OF SAWS

GENERAL MAINTENANCE

- Frequently give the saw a light coat of oil on the blade, to prevent rust. Keep the saw away from water.

- Keep the saw in the toolbox and be careful not to damage the blade by putting other tools on top of it.

- Keep the workbench uncluttered and be careful where you put the saw down.

- If the blade gets bent or buckled, straighten it at once.

THE ACTION OF THE TEETH

In order to properly maintain saws, it is necessary to understand how the teeth should look and how they work.

The ripsaw tooth resembles the cutting edge of the chisel. Its cutting edge strikes at practically a right angle to the wood fibres and the effect is as if a series of small chisels were set one behind another. Each tooth cuts out the full width of its edge and carries away the shaving (Fig. 1).

The teeth of the crosscut saw and the backsaw have a different shape, size and action from those of the ripsaw.

The form of the teeth is more like a series of knives that cut alternately on the two sides of the kerf (Fig. 2).

In sawing across the grain of the wood, the wood fibres must be cut on each side of the kerf so that a clean cut can be made. The teeth are therefore formed into sharp points on the outer side, so that they cut the fibres accurately.

The cut is started as the teeth make fine cuts, similar to the cuts of a knife, across the wood surface. Then as pressure is applied, the teeth go deeper and deeper, gradually bringing into action the full cutting edge of the teeth.

NOTES:

Fig. 1 — GULLET, 25 mm, 60°, ANGLE OF PITCH 0° - 3°

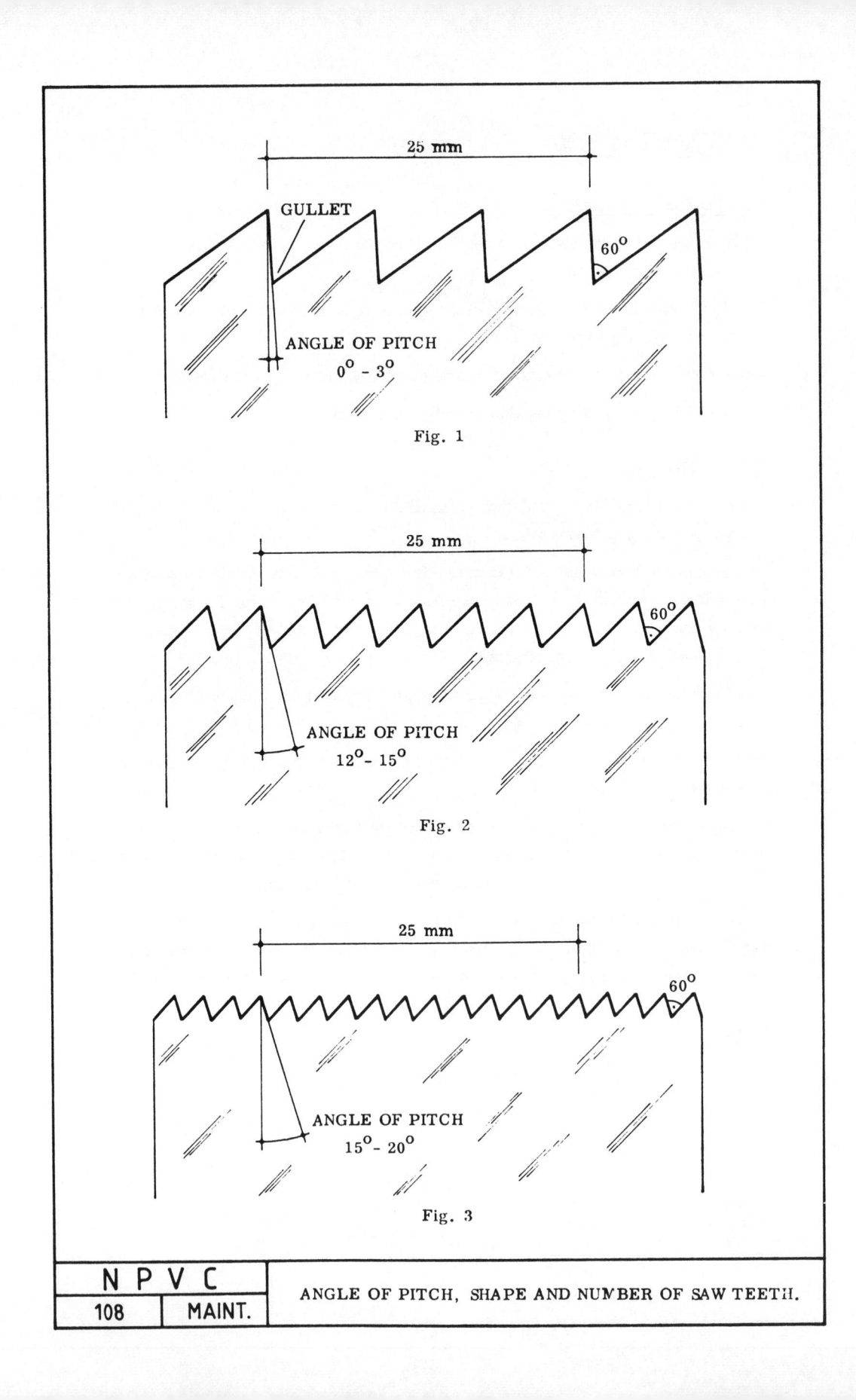

Fig. 2 — 25 mm, 60°, ANGLE OF PITCH 12° - 15°

Fig. 3 — 25 mm, 60°, ANGLE OF PITCH 15° - 20°

NPVC	ANGLE OF PITCH, SHAPE AND NUMBER OF SAW TEETH.
108 MAINT.	

ANGLE OF PITCH, SHAPE AND NUMBER OF SAW TEETH

The angle of pitch of the saw teeth is important in the maintenance of the saw. This angle is the measure of how far the face of the tooth is leaning from the vertical. The angle of pitch for the different saws is shown on the left page.

The smaller the angle of pitch, the faster the saw will cut and the more often it will need resharpening. A large angle of pitch means a longer life for the blade, while a small angle means more frequent resharpening and thus a shorter life.

THE RIPSAW

The blade of the ripsaw has 4 points (teeth) per 25 mm. The teeth of the ripsaw are rather big (Fig. 1).

The angle of pitch of these teeth is very small, from 0 to 3 degrees. This small angle means that the ripsaw will not cut properly across the grain, because the teeth will tend to tear the fibres. This saw is used only for cutting with the grain, where there is not any danger of tearing the fibres.

THE CROSSCUT SAW

The blade of this saw has 6 to 8 points per 25 mm, so its teeth are smaller than those of the ripsaw. The angle of pitch of the teeth is greater in order to get cleaner cuts; this also makes the work slower. The angle is between 12 and 15 degrees for this kind of saw (Fig. 2).

Do not use this saw for cutting along the grain; it is not designed for that and it will not guide as well as the ripsaw.

THE BACKSAW

The blade of the backsaw has 10 to 15 points per 25 mm, so the teeth are still smaller than those of the crosscut saw. The angle of pitch is between 15 and 20 degrees (Fig. 3).

NOTES:

Fig. 1

Fig. 2

Fig. 3

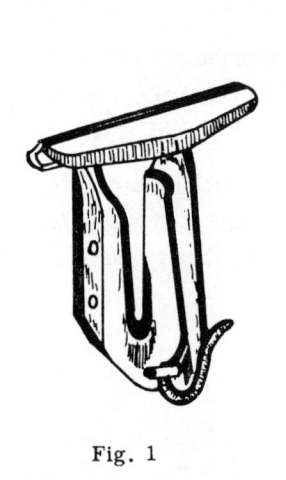

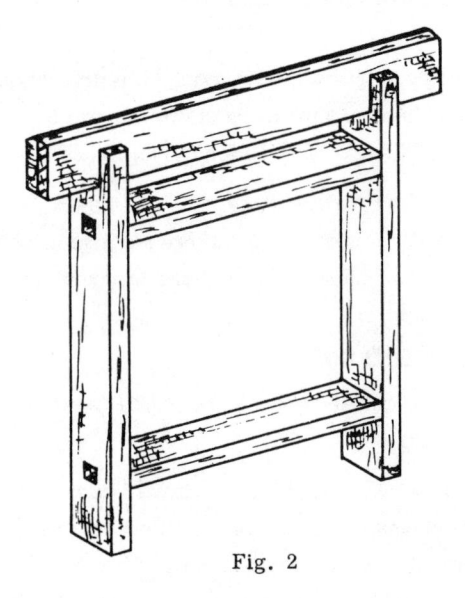

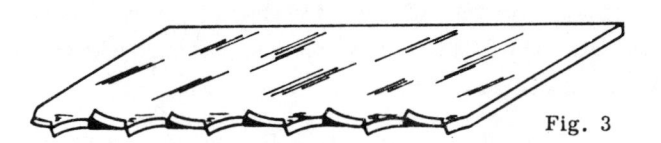

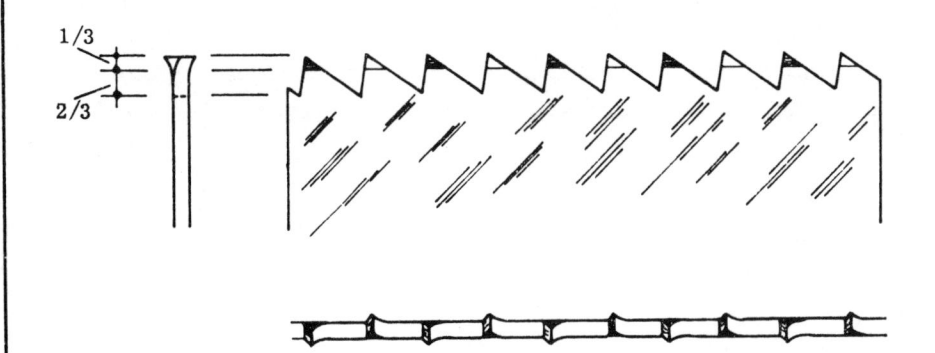

1/3
2/3

Fig. 4

N P V C		SETTING HANDSAWS.
110	MAINT.	

SETTING HANDSAWS

There are two main operations involved in sharpening a handsaw:

a. Setting
b. Filing

In order to make sure that the saw blade moves freely in the saw cut without any side friction, the saw teeth must be set. This is done by slightly bending the teeth at their tips (Fig. 3) to give more clearance in the kerf (saw cut). A saw should never be given more set than is necessary for the blade to move easily in its kerf. Too much set will cause the blade to wander out of line, too little set can cause the saw blade to buckle.

Only the tips of the teeth are set, never more than about 1/3 rd of the length of the tooth (Fig. 4). If you set the saw deeper than that you will buckle or crack the blade.

A saw need not be set every time it is filed, particularly if only a light filing is required. A saw can sometimes be filed two or three times before it needs to be reset and filed again.

- SAW VICE: During the maintenance of a saw, it must be securely fixed in position. A saw vice is used to hold it; either the wooden type (Fig. 2) called a horse, or the metal type (Fig. 1) which grips the blade more strongly but does not grip the whole length of the blade.

  Another type of wooden saw vice can be made by fixing the two wooden jaws to the handle of your toolbox.

NOTES:

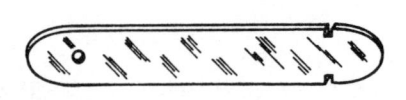

Fig. 1

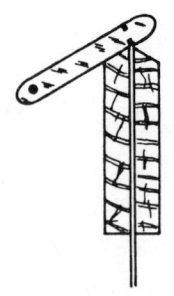

Fig. 2

Fig. 3

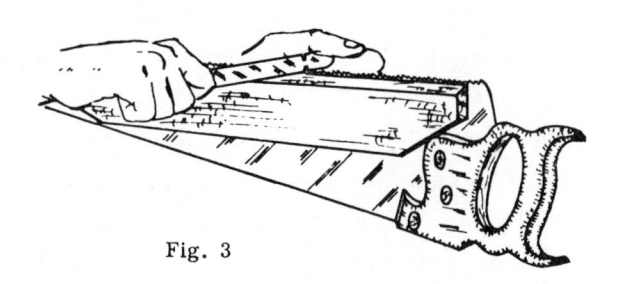

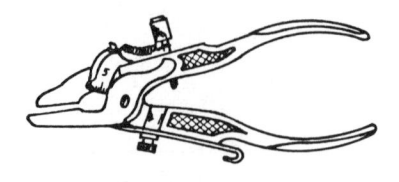

Fig. 4

Fig. 5

SETTING HANDSAWS.

HOW TO SET A SAW

To do the setting, we use a setting tool. This can be a strip of metal with some cuts in it, the cuts being the thickness of the different saw blades (Fig. 1). Setting has to be done by eye, and practice is needed to get the right bend. The top of the saw vice can be bevelled to act as a guide for the setting tool (Fig. 2).

- Hold the saw firmly in a good position. When setting the first tooth, take care that you bend it to the same side that it was bent before.

- Start at the heel of the blade and bend every second tooth. When you finish one side, turn the blade around and do the other teeth (Fig. 3).

- It is important that the set is exactly the same on each side, otherwise the saw will run (saw out of line).

Instead of the setting tool mentioned above, pincer type sawsets (Figs. 4 & 5) are often used. You simply place the set over the tooth and squeeze the handles. These sawsets are adjustable, so when you use an unfamiliar one, it is best to set a few teeth and examine them before you set the whole saw. If the teeth are set too much or not enough, you can then adjust the sawset accordingly.

SETTING THE RIPSAW

When the saw cuts with the grain, like the ripsaw, the fibres of the wood don't tend to move back into the cut, so the ripsaw doesn't need a big set to have enough clearance between the blade and the kerf.

SETTING CROSSCUT AND BACKSAWS

Crosscut fibres tend to move back into the kerf, so these saws need a bigger set to get enough clearance.

It is important to note that the wider setting of these saws makes them unsuitable for cutting with the grain, because the kerf will be too wide to guide the saw.

NOTES:

Fig. 1

Fig. 2

Fig. 3

WRONG

RIGHT

Fig. 4

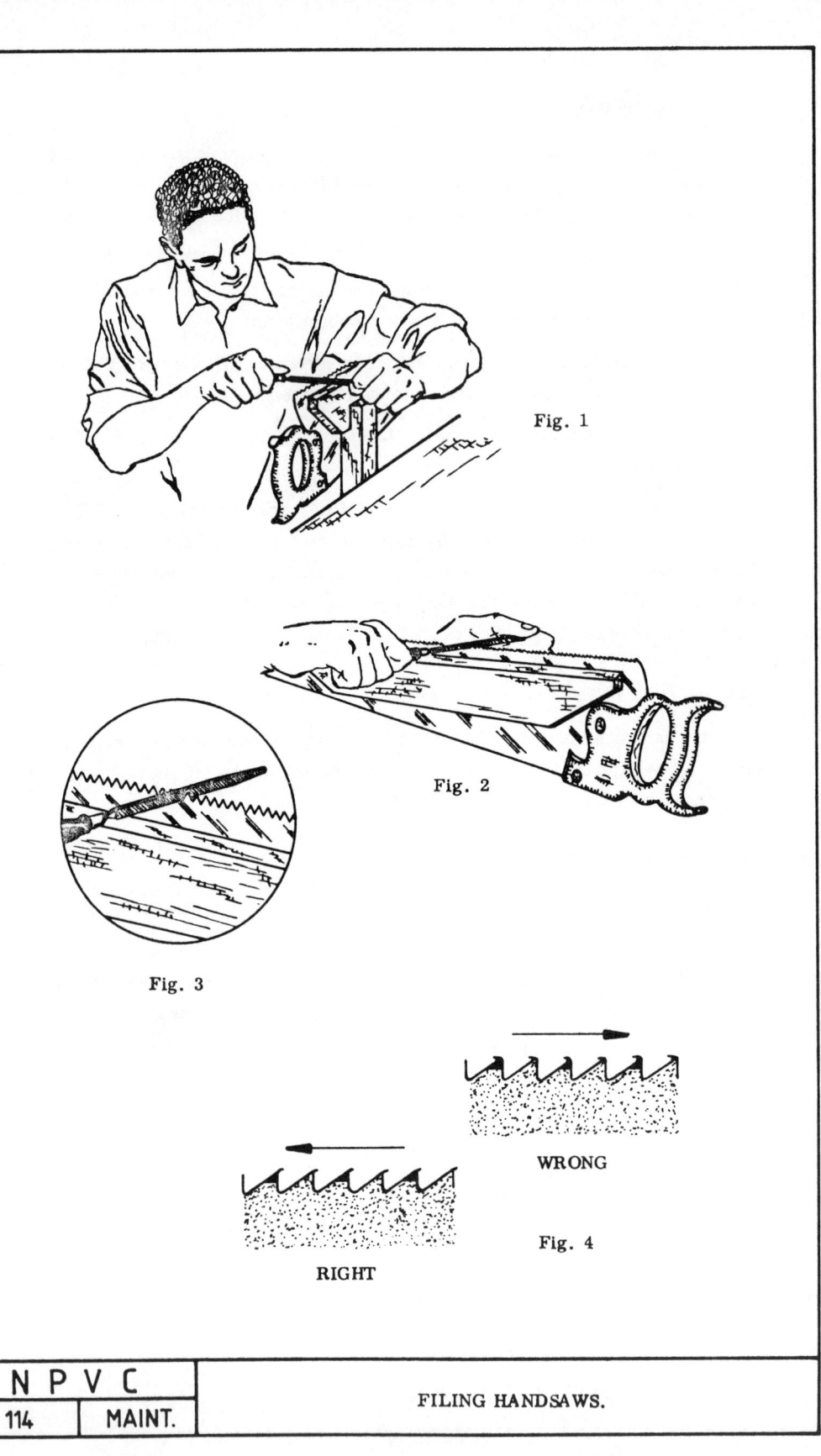

N P V C	
114	MAINT.

FILING HANDSAWS.

FILING HANDSAWS

After the saw has been set it can be filed.

Secure the saw in a saw vice, with the teeth sticking out just a little way from the vice jaws. If the teeth stick out too far, the filing will cause a screeching sound.

The top of the saw vice should be at about the level of your armpits, or slightly below (Fig. 1). While filing, you must be able to constantly check the shape of the teeth and this rather high position enables you to see them properly.

In order to avoid eyestrain and ensure a good job of filing, it is essential to have good light. Work in front of an open window if possible, so that the light shines on the saw teeth.

USING THE SAWFILE

Small metalworking files with a triangular cross section are used for filing saws. The following points are important to remember in using the saw file:

- Hold the file handle in your right hand (Fig. 2).

- Hold the tip of the file gently between your thumb and forefinger of the left hand (Fig. 2).

- Exert pressure on the forward stroke only.

- Make long slow cutting strokes, not short fast ones.

- Keep the file level (Fig. 3).

- Use enough pressure to make the file cut, but no more.

During filing, a small burr is formed at the tip of the teeth. This burr can improve the cutting action of the saw when it is filed in the right direction. Therefore the saw must always be filed from toe to heel (Fig. 4).

NOTES:

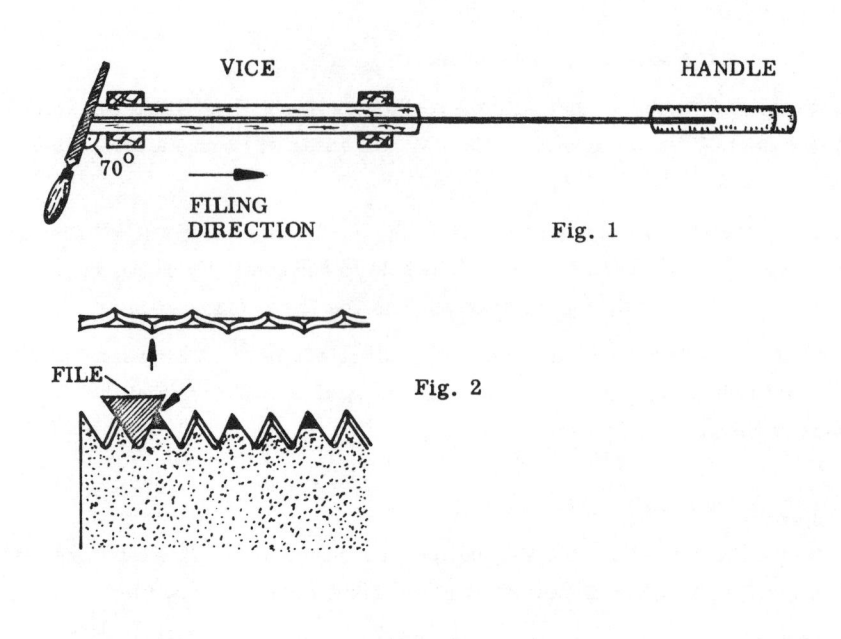

VICE — HANDLE

70° FILING DIRECTION → Fig. 1

FILE

Fig. 2

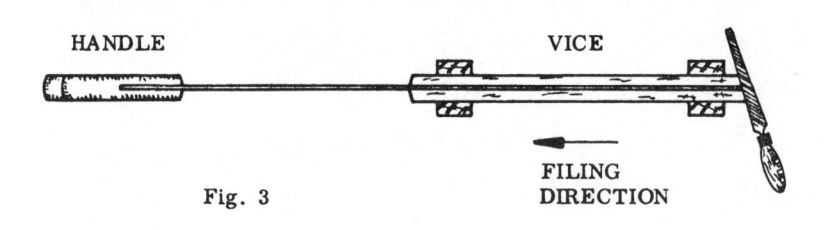

HANDLE — VICE

Fig. 3 ← FILING DIRECTION

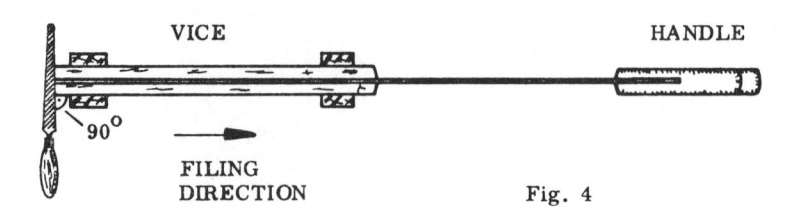

VICE — HANDLE

90° FILING DIRECTION → Fig. 4

NPVC 116 MAINT.

FILING HANDSAWS.

HOW TO FILE THE CROSSCUT SAW

Study Fig. 1 very carefully to understand the proper starting position for filing this saw. Work according to the following sequence:

1. Place the toe of the saw in the saw vice with the handle to your right.
2. Find the first tooth in the toe that is bent towards you. Place the file in the first gullet (V-notch between the teeth) to the left of that tooth (Fig. 2, arrow).
3. Hold the file across the saw blade at an angle of about 70 degrees, with the point towards the saw handle.
4. If the teeth are of the proper shape (see section on Angles of Pitch) press the file gently into the gullet and let it find its own placement against the two teeth (Fig. 2). Remember to keep the file level.
5. Push the file forward, cutting the front surface of one tooth and the back surface of another.
6. Release the pressure on the file during the backstroke.
7. File every other gullet until you are about halfway through the saw, then make a pencil mark at this point.
8. Shift the blade in the vice until the head end is held more securely and continue filing every other gullet until you reach the handle.
9. When every other gullet has been filed from one side, turn the saw around in the saw vice so the handle is to the left (Fig. 3).
10. Find the first tooth in the toe of the saw that is bent towards you and put your file into the gullet to the right of that tooth.
11. Hold the file at an angle of about 70 degrees across the blade, with the point towards the saw handle.
12. File every other gullet as before, until you reach the saw handle.

- REMEMBER:
  - Inspect your work frequently to make sure that you are getting the teeth properly shaped. Remember that the angle of pitch must be correct.
  - Throughout the filing, give each tooth the same number of strokes with the file. This helps keep the teeth all the same size and shape.
  - If you get one tooth out of shape, don't be too concerned; it can be left as it is.

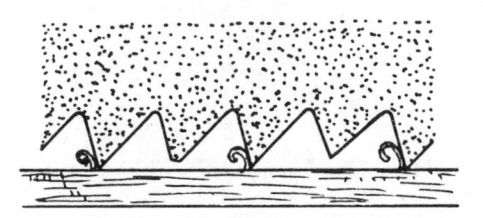

Fig. 1

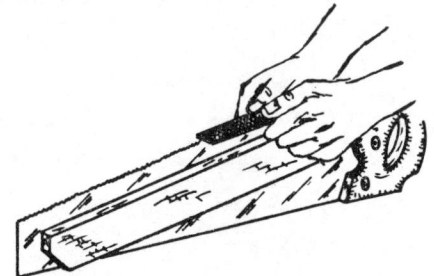

Fig. 2

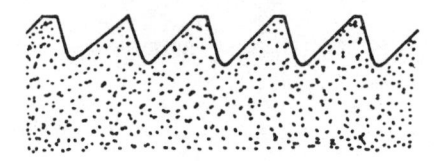

Fig. 3

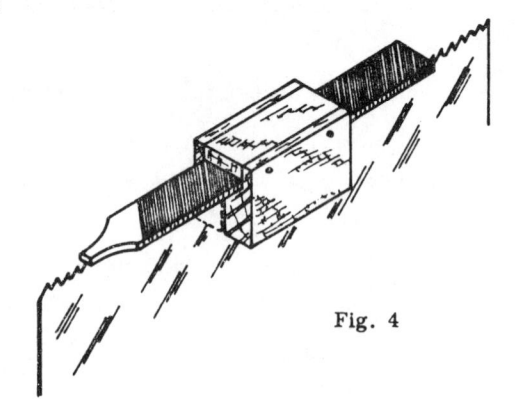

Fig. 4

FILING HANDSAWS & TOPPING A HANDSAW.

- Turn the file occasionally, so that it is used evenly.
- If you think you have lost your place or skipped a gullet, look for the last shiny tooth in the light.

HOW TO FILE THE BACKSAW

This saw is filed in the same way as the crosscut saw, except that the angle of pitch is different.

Be careful with the pressure you apply with the file, because the teeth are very small and are easily filed out of shape.

HOW TO FILE THE RIPSAW

The same general procedure is used for filing ripsaws as for crosscut saws. There are two important differences:

1. The angle of pitch is different.
2. The file is held at 90 degrees, not 70 degrees, to the blade (Fig. 4, previous page).

(Some people file the ripsaw from one direction only, filing all the teeth at once instead of filing every other tooth and reversing the saw).

TOPPING A HANDSAW

If you always file the saw correctly, without twisting the file or filing some teeth more than others, the teeth will always have the correct form and equal height.

When you make small mistakes in filing, or use the saw carelessly, the teeth will become out of line, different in height and irregular in shape. The result is that only some of the saw teeth can cut; the others don't touch the wood (Fig. 1). Topping is done to correct this problem.

WELL KEPT AND MAINTAINED SAWS NEVER NEED TOPPING!

HOW TO TOP A SAW

Topping must always be done before you set or file the saw.

- To top a saw, you run a flat metalworking file over the ends of the teeth, moving along the length of the saw. Be very careful to keep the file square to the saw blade and flat.

- One way to do this is to grasp the file in both hands by the edges, thumbs on top and the fingers under, touching the saw blade and guiding the file (Fig. 2, previous page).

- A wooden block may also be used to hold the file in the proper position (Fig. 4, previous page).

- File until there is a small shiny point on each tooth. When you have made two or three light strokes and there are still some teeth that have not been touched (Fig. 3, previous page), don't keep on filing. Too much topping will make it difficult to reshape the teeth.

- Next, file the teeth to the correct form again. This is called reshaping the teeth. When all the teeth have the same height and shape they are ready to be set and filed again after setting, to sharpen them.

NOTES:

PART 3: RURAL BUILDING MATERIALS

Man learned early to protect himself from the weather as well as from his enemies by making shelters of various kinds.

The type of structure he made depended on the locally available building materials, and on whether the shelter was meant to be permanent or not. Compare the sketches below, for example.

The extreme northwest of Ghana provides the inhabitants with such universally used building materials as sand, mud, stone, laterite, timber, grass and straw. However, the traditional way of building with these materials does not always produce a very permanent structure; with the result that almost every dry season the houses must be thoroughly maintained, repaired and even rebuilt. In order to make more durable structures, the Rural Builder should include in his technical knowledge the uses of local and modern materials, and of binding materials such as lime or cement. Building materials are described in the following chapters.

TENT

STRAW HUT

MUD HUT

N P V C	
122	MAT.

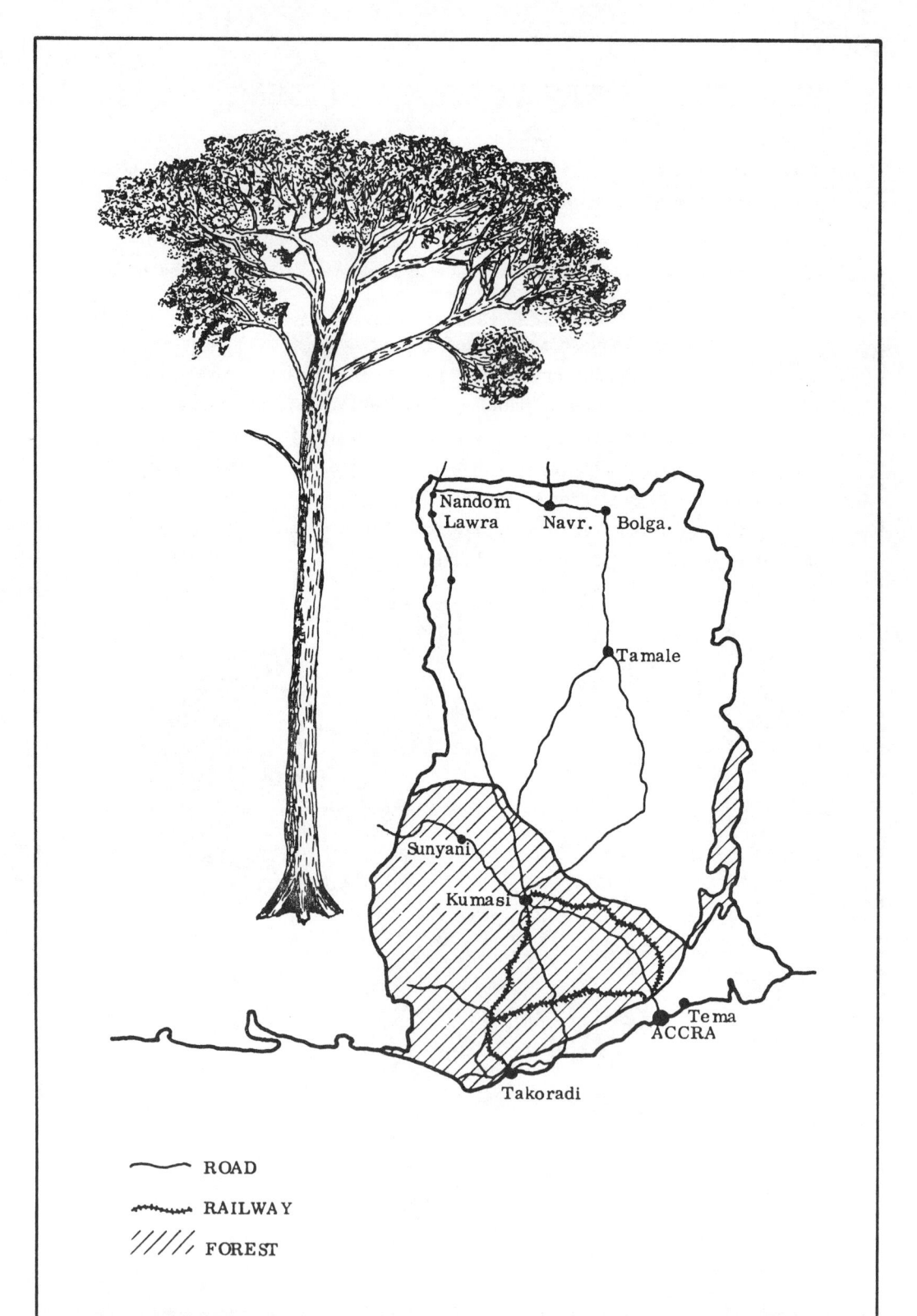

ABOUT WOOD IN GENERAL.

- ROAD
- RAILWAY
- FOREST

Nandom, Lawra, Navr., Bolga., Tamale, Sunyani, Kumasi, Takoradi, Tema, ACCRA

ABOUT WOOD IN GENERAL

In Rural Building we work a great deal with wood. Wood has always served man for many different purposes: for tools (handles of hoes, knives and axes), for weapons (bows and arrows), for housing (ladders, doors and windows), furniture and firewood.

For traditional building purposes, nearby trees and bushes are cut and the branches as well as the trunks are used.

In Rural Building, we work with sawn timber. Sawn timber is cut from the trunks of very tall trees. Such trees are not common in the northern parts of Ghana. They are found in the rainforest areas of the south, where they are felled (cut down), sawn into boards and transported to the north and elsewhere.

The most common woods we use in northern Ghana are Wawa and Odum. Others, like Emire or Mahogany, are sometimes used.

There are 85.000 square kilometres of forest in Ghana, from which comes more than 200 different species of wood. Most of the wood that is cut is not used in Ghana but is sold to other countries to bring in money. It is one of the largest sources of income for Ghana. The principal kinds of wood that Ghana exports are Afromosia, Wawa, Utile, Sapele, Odum and Mahogany.

To make sure that these woods are available in the future, the cutting is controlled by Forestry Acts and efforts are made to reforest the cut areas. These new trees will not be ready to cut for a long time.

The map on the left page shows the areas where the trees are cut, the locations of the roads and railways that bring the wood into the harbours at Takoradi and Tema and the roads by which wood is exported to neighbouring countries. It is transported either as whole logs or as sawn timber and also as timber products made from waste wood or wood chips, like chip board or plywood. The mills that convert logs to sawn timber and timber products are located all over the forest.

THE STRUCTURE AND GROWTH OF THE TREE

A tree has three main sections: the roots, the trunk and the crown. The crown is made up of the branches, twigs and leaves (Fig. 1, next page).

In order to grow, the tree must have water and minerals from the soil. The water and minerals are taken out of the soil by the roots and brought to the

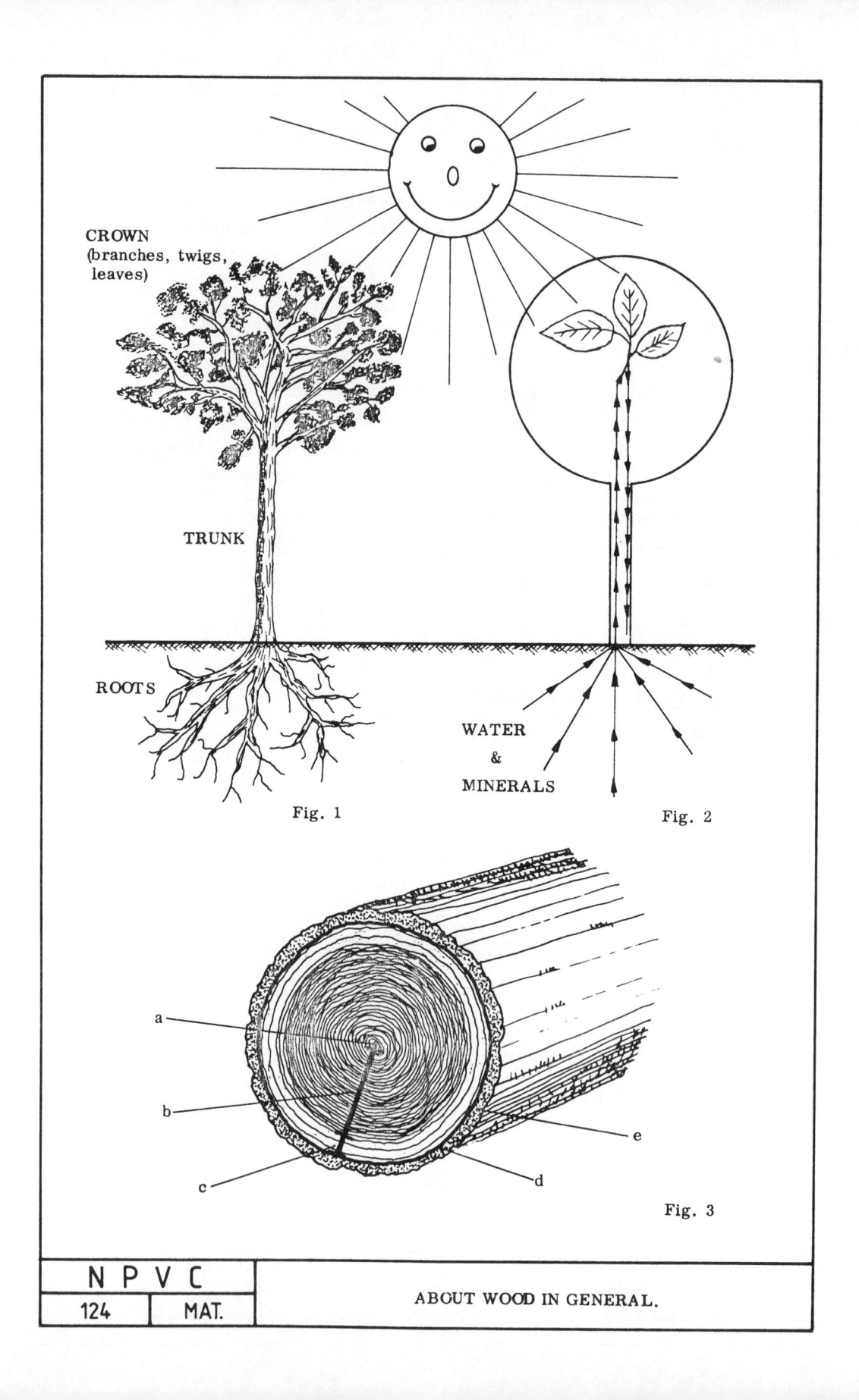

CROWN
(branches, twigs, leaves)

TRUNK

ROOTS

Fig. 1

WATER & MINERALS

Fig. 2

a
b
c
d
e

Fig. 3

N P V C	
124	MAT.

ABOUT WOOD IN GENERAL.

leaves through the outer layers of the sapwood (Fig. 2).

In the leaves, the sunlight acts on the water, minerals and air to make the materials that the tree uses to build new roots, wood and leaves.

The material made by the leaves is brought down to the roots and growing parts of the tree by the inner layers of the bark. If the bark is damaged by cutting or bush fire, the sap can't move up and down and the tree might die.

HARD AND SOFT WOOD

We talk about two main categories of wood: hardwoods and softwoods. These categories are only trade terms, they do not indicate that the wood itself is either hard or soft. The difference comes from the way of growth of the tree. Almost all the trees that grow in Ghana are hardwoods, but the actual wood itself may be hard or soft in character. For example Wawa, which is classed as a hardwood, is actually very soft in character and easily worked.

THE STRUCTURE OF WOOD

The tree trunk is made up of five layers (Fig. 3).

The pith (a) is the centre or heart of the tree.

The heartwood (b) is the fully developed mature wood which surrounds the pith. It is usually dark in colour and hard. The heartwood does not play an active part in the growth of the tree; it is only for strength, to support the tree.

The sapwood (c) is the layer around the heartwood which is lighter in colour and softer. Sapwood is immature wood, it will harden and darken and become heartwood as the tree grows. A new ring of sapwood is added every year on the outside of the older wood.

Sapwood should not be used for construction purposes, because it is soft and easily attacked by termites. It is always removed when wood is cut for construction purposes.

The cambium (d) is a soft, greenish layer between the sapwood and the bark. It is the "factory" of the tree, making new wood at the inside and new bark at the outer side. The cambium is fed by materials brought from the leaves by the inner layers of the bark.

The bark (e) protects the cambium from cold, fires, insects and animals, and it also transports food material from the leaves to the cambium and to the roots.

The annual rings are the rings you see (Fig. 3) in the heartwood and sapwood.

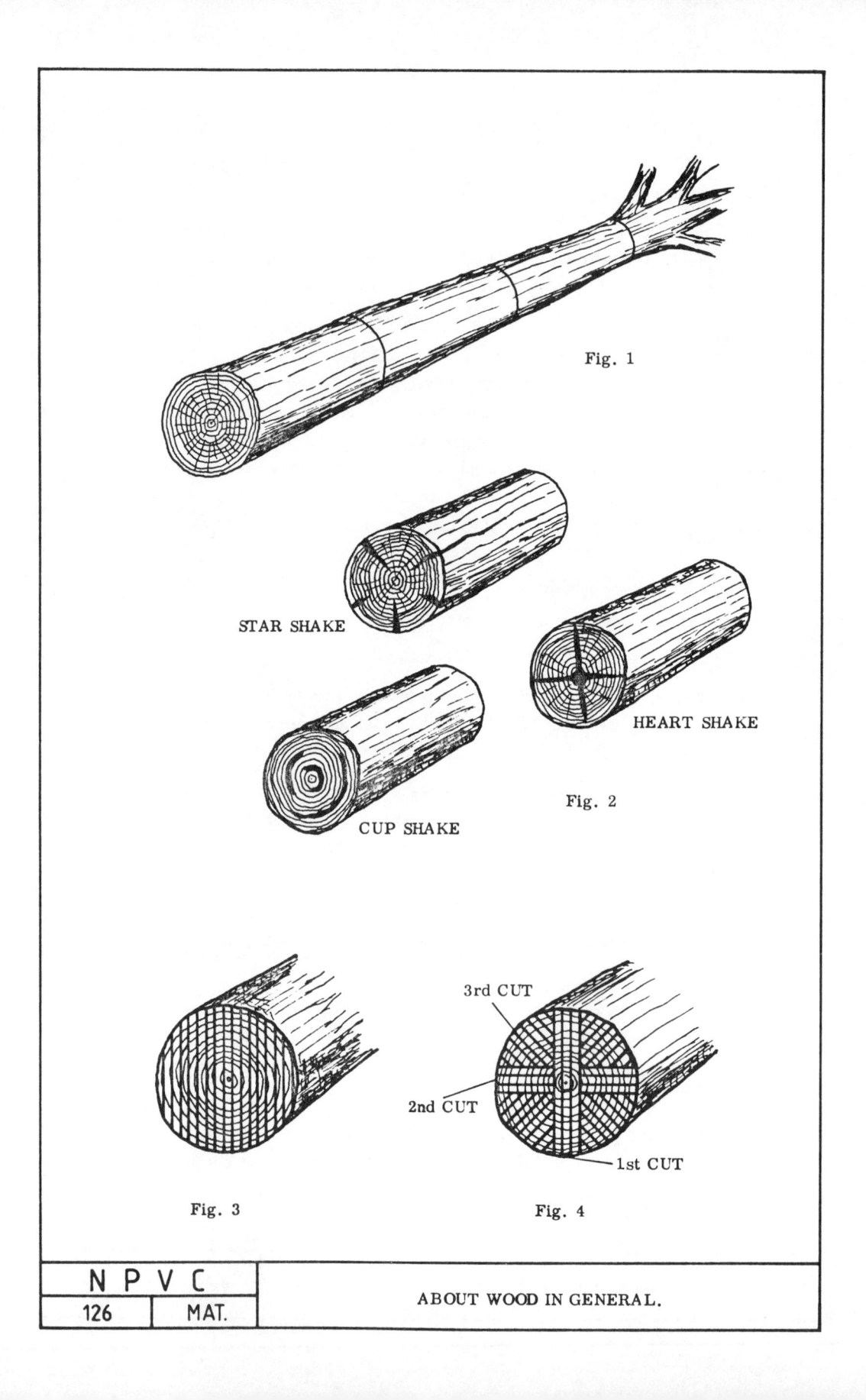

Fig. 1

STAR SHAKE

HEART SHAKE

CUP SHAKE

Fig. 2

Fig. 3

3rd CUT
2nd CUT
1st CUT

Fig. 4

N P V C		ABOUT WOOD IN GENERAL.
126	MAT.	

These are formed by the growth of the tree, by the layer of new wood that is added each year.

The tree grows all year long but in the rainy season, when it gets more water, it grows faster than in the dry season. The new wood formed in the rainy season is lighter in colour than the wood formed in the dry season, so they appear as light and dark rings in the wood. One light ring and one dark ring are formed each year; together they make an annual ring (annual means every year).

Don't forget: HARD WOOD is not the same as HARDWOOD is not the same as HEARTWOOD; they all mean different things.

THE PATH FROM STANDING TREE TO SAWN TIMBER

After the tree is felled, the crown is removed and the bark is taken off so no insects can hide in it. The trunk is cut into logs up to 12 metres long (Fig. 1) for transport and handling purposes. These logs are brought to the sawmill where they are converted to sawn timber. Converting the logs to timber means sawing them into boards, planks etc., which can be used for construction purposes.

If the logs cannot be converted immediately, they are kept in water to prevent the formation of shakes. These are long cracks in the unsawn log (Fig. 2). The most common ones are: cup shakes, which occur when the annual rings fail to grow together and star and heart shakes, which can occur in the growing tree or in the cut log as it dries.

There are different ways of converting logs. The method which is chosen depends on the thickness and species of the tree and the quality of the wood that is needed.

Plain sawing is the easiest, cheapest and most common way of converting tropical woods. It is also called the "through and through" method.

The log saw starts cutting from one side and continues, cutting off one board after another until the whole log is converted to timber (Fig. 3).

The second method, quarter sawing, requires better equipment and more work than plain sawing. With this method, almost all of the annual rings will be square to the surface of the board, which we will see later is an advantage over plain sawn boards (Fig. 4).

There are still other methods of conversion, but they are not important for us in this course.

CONVERSION TERMS FOR SOLID TIMBER

- Scantlings are pieces about 75 mm wide by 50 mm thick.

- Boards are pieces more than 150 mm wide and less than 50 mm thick.

- Planks are pieces above 200 mm wide and between 38 and 100 mm thick.

- Baulks are more square shaped pieces, about 100 mm wide and 50 mm or more thick.

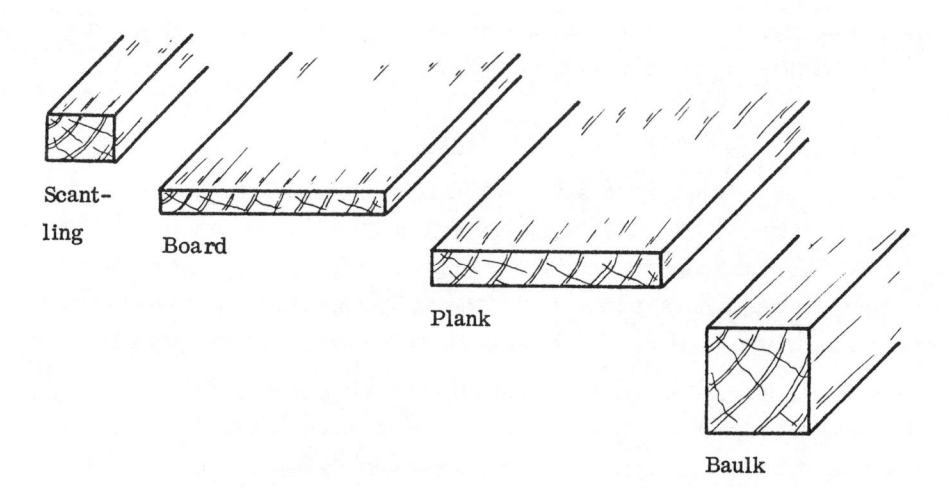

HOW TO ORDER TIMBER

Before you place an order for wood, make sure that you list the correct sizes, quantities and kinds of wood.

Boards are sold in different measurements. Very often the sizes are still given in the imperial system, but the metric system is becoming more common everywhere.

The surface of sawn timber is still rough, so you have to allow for planing the boards when you order timber (refer to Rural Building, Basic Knowledge, p. 90).

Example:

		Size		
No.	Kind of wood	Thickness	Width	Length
50	Odum	2,5 cm	30 cm	2,5 m up
40	Wawa	3,8 cm	30 cm	3,0 m up

If a minimum length of sawn timber is required, the word "up" is added. So "3,0 m up" would mean 3,0 m and longer.

THE PROPERTIES OF WOOD

To be able to use timber properly and store it in the correct way we need to have some knowledge about:

- the moisture content of wood, and
- wood shrinkage.

THE MOISTURE CONTENT OF WOOD

When a tree is cut down, the trunk still contains a large amount of water which has been stopped on its path through the trunk and remains trapped there. The weight of the water at this time is between 1/3rd to 1/2 the weight of the tree.

Thus the moisture content (amount of water remaining in the wood) is very high when the tree is freshly cut. Some of the water evaporates (dries off) as the logs are brought to the mill and more will evaporate after the logs are sawn into timber. Each time the moisture content changes, the size of the timber also changes.

SHRINKAGE

As the wood dries, it becomes smaller in size. This is what we mean by shrinkage. Because of the wood structure, it does not shrink equally in each dimension; shrinkage will be up to 10% along the annual rings, about 5% at right angles to the rings and no more than 0,5% in the length of the board.

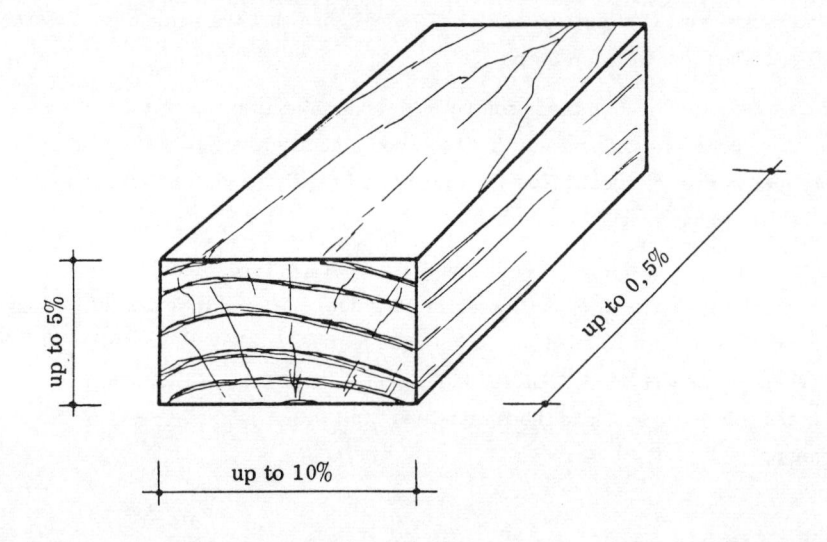

Here we can see the advantage of quarter-sawn timber, in which the annual rings are mostly at right angles to the width of the board. The biggest size change (10%) will be in the thickness of the board, where it usually doesn't matter so much. The shrinkage in the width will however still be up to 5%.

In plain-sawn boards, the rings lie mostly across the width of the board and the board will shrink more and also tend to get out of shape as it dries, because the rings change directions within the board.

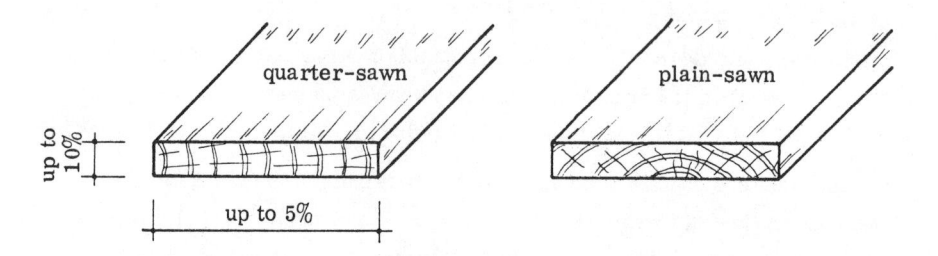

Shrinkage in the length of the timber can almost be ignored, and will only matter when you join boards together end to end over longer distances.

You should know that different woods shrink differently, for example Odum will shrink more than Wawa.

The weather has a great effect on the size of wood. In northern Ghana especially, the very great changes in the humidity (the moisture in the air) between the dry and rainy seasons result in a lot of problems for the builder.

The very dry air of the dry season and especially that of the harmattan causes wood to dry out too much and shrink. You will find that the boards easily crack or split when you work on them.

During the rainy season the humidity is very high and this results in a higher wood moisture content. This means that the wood actually expands quite a bit. The wood may feel wet to the touch. You might plane a board straight and true one day, and find the next day that it is bent again.

Keep the effects of the weather in mind when you construct anything out of wood. For instance, you want to build a solid door out of Odum. If you build it in the dry season it should not fit tightly into the door frame, so that it can still open when it swells in the wet season. If you build it in the wet season it should fit well, so that the gap between it and the frame is not too wide when it shrinks in the dry season.

THE PROPERTIES OF WOOD.

NATURAL SEASONING

The builder should know that he must never use freshly sawn timber. The timber must first be seasoned, which means that it is dried to a certain moisture content which is most suitable for building work, so that it changes its size as little as possible after it is used in a piece of work.

The kind of seasoning we do in Rural Building is "natural seasoning", and it is done by storing the wood for some months.

After seasoning, the moisture content of the wood should be low enough to use it for building work. However, even well seasoned timber will still be affected due to the changes in humidity from the dry to rainy seasons.

Well seasoned timber will still shrink or expand up to 6% in the direction of the annual rings. For example, a board which is sawn in such a way that the annual rings are along the width (see plain sawing) might have a width of 30 cm during the dry season, while in the rainy season it can expand up to 31 1/2 cm wide. This means that whenever possible you should choose your boards in such a way that the changes in size don't cause problems in the finished piece.

NOTES:

WARPING

Seasoning can cause boards to bend. This happens because the moisture content differs from one part of the board to another, especially if one part gets more sun or rain, when it is stored improperly. The boards should be stored in such a way that air can reach every side of them and all sides are equally dry.

Wood that is not straight grained also tends to warp as it dries.

There are four different kinds of warping: bowing, cupping, springing and twisting. These are illustrated below.

Warping of boards can be partly avoided by proper storage of the wood, which we will discuss in the next section. To prevent warping of finished workpieces such as door and window frames, they should be installed as soon as possible after they are completed.

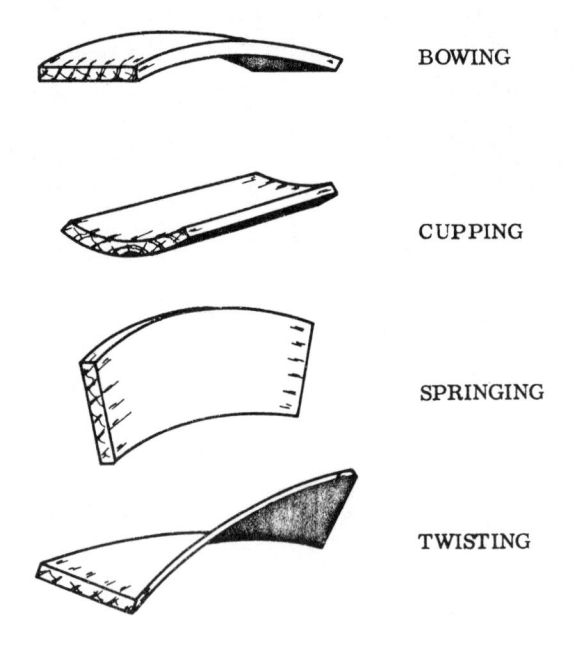

BOWING

CUPPING

SPRINGING

TWISTING

TIMBER PILING

In previous lessons, we learned about why seasoning is important to reduce shrinkage. Seasoning also helps to prevent decay and attack by insects and fungi. In natural or air seasoning, the wood is kept protected from sun, rain, and insects, but air is permitted to circulate freely around the wood.

The site where we put the wood to season is very important. It must be open and well drained; all weeds and grass should be removed and the base should be covered with gravel, or even better, with concrete, to prevent growth of new weeds under the stack. Ashes can also be spread around to keep away termites.

Strict cleanliness should be observed around the stacks. Sawdust and short pieces of wood left lying around can start rotting and attract termites or fungi, or they might even catch fire during the dry season and destroy the entire stock of wood. Take extra care with Wawa, as it is easily attacked by insects or fungi.

The entire pile should be shaded from the sun, to keep the timber from drying out too rapidly. The stack should be oriented with the length in the east-west direction, to minimize the effects of rain and sun.

LAYOUT OF THE WOOD STACK

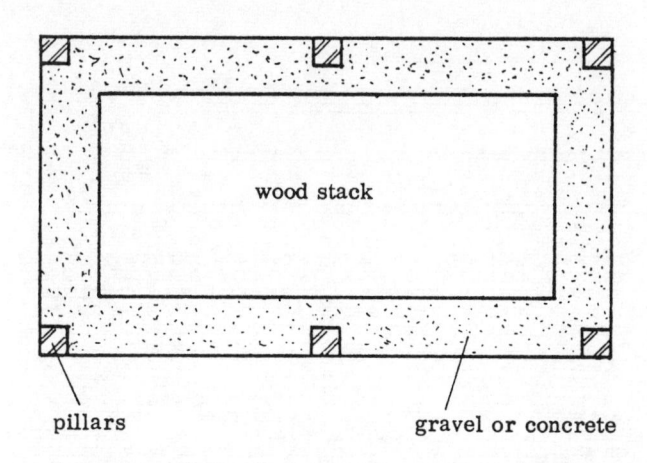

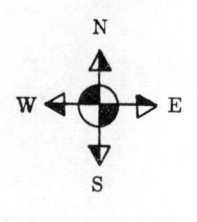

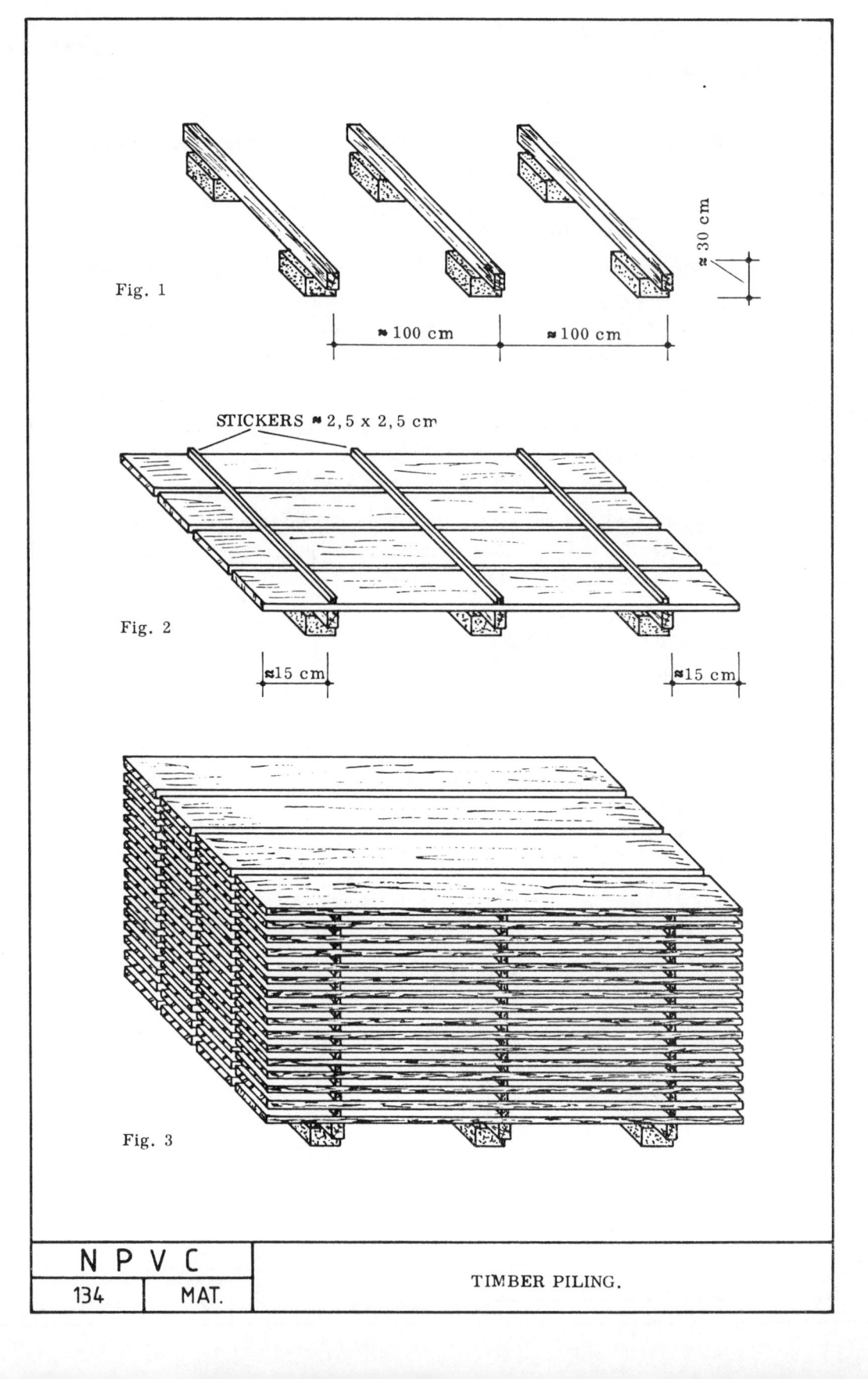

Fig. 1

≈ 30 cm

≈ 100 cm ≈ 100 cm

STICKERS ≈ 2,5 x 2,5 cm

Fig. 2

≈ 15 cm   ≈ 15 cm

Fig. 3

N	P	V	C
134	MAT.		

TIMBER PILING.

MAKING THE STACKS

The bases of the stacks are sandcrete blocks (not wood, which might be attacked by termites). On these blocks we set the cross pieces, which are straight baulks without any twisting, because they must carry the whole weight of the stack. The length of the cross pieces will be the width of the stack. They should be perfectly in line on top of the bases (Fig. 1).

The pile should be level from side to side, but may slope a bit lengthwise. Place the cross pieces about 100 cm apart. The end pieces should be about 15 cm from the ends of the boards (Fig. 2).

On top of the cross pieces, put the first layer of boards. Between the layers of boards, put wood strips (stickers) to allow air to flow between the layers (Fig. 2). The stickers should be all exactly above the cross pieces and each other as we build up the stack (Fig. 3). If this is not done correctly the boards may start bending.

The stickers should be square in section so you don't have to take care to put them all flat or all edgeways. They should be around 25 mm thick. All the stickers should have the same thickness, or they may cause the boards to bend or the stack to collapse.

Always stack the wood in such a way that the pieces you will need first are on top.

If you have some wood that you want to reserve for a special purpose, it may be secured by tacking a short stick across the end of the pile.

The stack should always be covered. Make sure that you leave the stack covered and in order after you take out wood.

NOTES:

TIMBER PILING.

SPECIFICATIONS OF WOODS WIDELY USED IN NORTHERN GHANA

ODUM

Local trade name: Odum

Standard (international) trade name: Iroko

The tree is up to 60 m high and 7 m in girth (circumference).

Wood characteristics:

The wood is hard and of medium weight.

It is variable in colour, from yellow to dark brown. When freshly felled, it is yellowish green.

The sapwood is paler, about 2,5 to 7,5 cm wide and it is quite distinct from the heartwood.

Durability:

Odum is very resistant to decay when kept dry but it is liable to be attacked by fungi in damp situations. The sapwood is often attacked by pin-hole borers and termites.

Seasoning:

It is easily naturally seasoned.

Working qualities:

It can be worked with moderate ease by most hand tools and it finishes well. Stonelike deposits are sometimes present and these may cause damage to cutting edges. The wood can be nailed, screwed or glued with no problems. When very dry it can be difficult to nail.

Uses:

Odum is probably the most generally useful tropical African hardwood and it is widely used locally for all kinds of construction work and carpentry.

WAWA

Local trade name: Wawa
Standard trade name: Obeche

The tree is up to 55 m high and 5,5 m in girth.

Wood characteristics:

The wood is soft and light in weight. It is nearly white to pale yellow in colour and there is no clear distinction between the sapwood and heartwood. The sapwood is about 7,5 to 10 cm wide.

Durability:

Wawa is not resistant to decay. Seasoned timber is liable to be attacked by powder post beetles and termites.

Seasoning:

Natural seasoning is rapid and satisfactory, with only very little warping and little inclination to split.

Working qualities:

It works easily with all hand tools. To avoid roughening the surface, use very sharp tools. The wood is rather soft and takes nails and screws easily, but it does not hold them well under hard use.

Uses:

It is used for formwork, scaffolding and furniture, where it is not exposed to attack by termites. It can be used only where it will be protected against moisture and rain.

NOTES:

Fig. 1

Fig. 2

Fig. 3

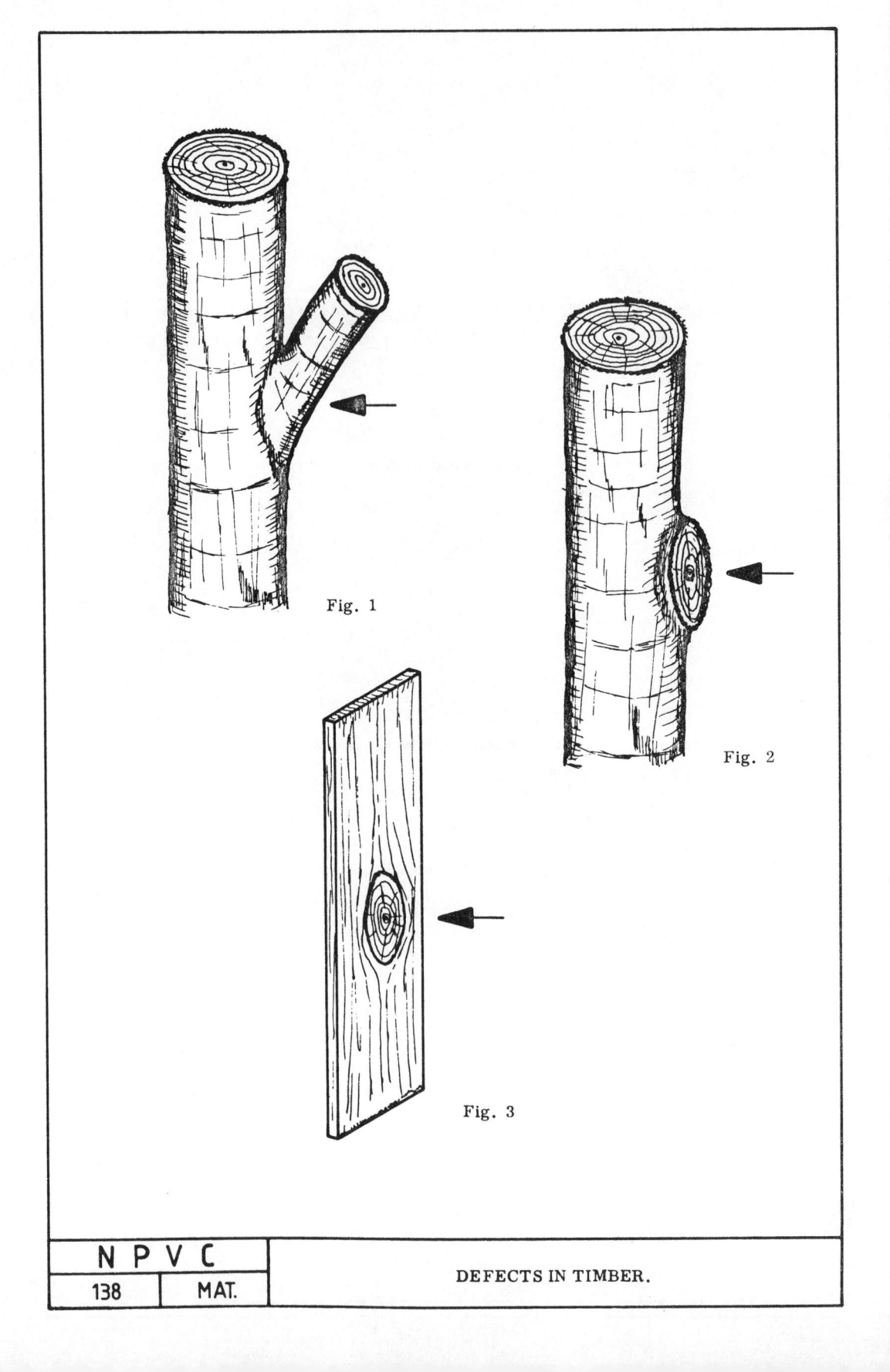

N P V C		DEFECTS IN TIMBER.
138	MAT.	

DEFECTS IN TIMBER

In order to choose the right timber for the work, we need to have some knowledge about defects and diseases in timber. A defect is an irregularity or weakness in the wood which reduces its usefulness and suitability.

The common defects in timber are: knots, twisted grain, checks, wane or waney edge and deadwood.

KNOTS

The place in a tree trunk from which a branch has grown out is called a knot. Each knot marks the junction of a branch with the stem.

There are "live" and "dead" knots.

When a branch is broken off or damaged, a small piece is left attached to the tree. The tree continues to grow around the branch piece, eventually burying it in new wood. These dead pieces of branches are known as dead knots. They have no connection to the living wood, but they occupy a place in the tree, with living wood surrounding them. When the tree is converted to timber, these knots often fall out.

When a tree is felled, all the branches along the stem will be cut off (Figs. 1 & 2). They will leave a knot which is called a live knot (Fig. 3), because it comes from a living branch. Live knots are sound, healthy knots and are always firmly fixed in the wood.

Knots are more or less common in all timber. As long as they remain in place, the presence of a few knots will not harm a piece of timber. However, knots also weaken the wood in some ways and pieces with many knots should not be used for parts which carry heavy loads.

Trees grown in the forest are usually tall, with all the branches at the crown and not along the stem where they could leave knots. Trees which stand apart from other trees tend to have more branches lower on the stem, forming knots which appear when the log is converted.

NOTES:

TWISTED GRAIN

This defect occurs when the tree grows crookedly. The sawn timber tends to twist and it is difficult to plane and chisel because of the changing direction of the grain.

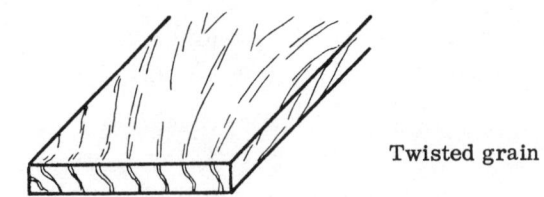

Twisted grain

CHECKS

Splits which occur during the seasoning of the wood are known as checks.

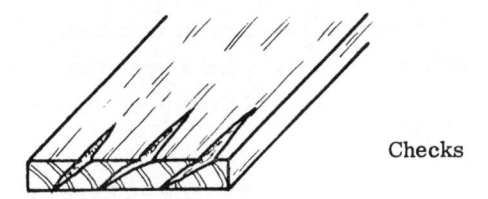

Checks

WANE OR WANEY EDGE

This defect is due to a lack of wood on the edge of the timber, from whatever cause.

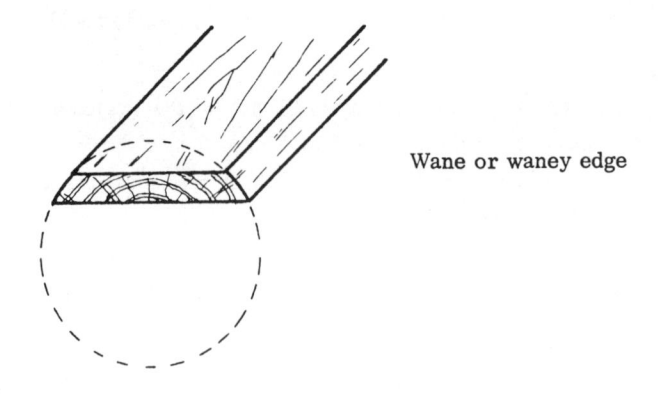

Wane or waney edge

DEADWOOD

Timber made from dead standing trees is called deadwood.

DEFECTS IN TIMBER.

DISEASES IN TIMBER

Diseases in building timber are caused mainly by attacks from fungi and insects

FUNGAL DISEASES

A fungus is a kind of plant which is not able to make its own food from sunlight, air and water, as most plants make theirs. Instead, it must get food by breaking down dead matter such as wood.

The best places for fungi to grow are dark, damp, warm places with little air movement.

There are many kinds of fungal attack on wood. We will only deal with the most common ones, dry rot and blue stain.

Dry rot is the most common fungal disease of building timber. It spreads easily and since the fungus actually feeds on the wood, it can cause a tremendous amount of damage. Dry rot is especially a problem with built-in work such as frames or cupboards where the wood is in direct contact with masonry work, because there it is often damp.

The appearance of the infected timber depends on the age and the extent of the disease. In the early stages, it looks as if pieces of thread are hanging from the wood. These quickly develop into a network that looks like a spiderweb, gray in colour. If the wood is very damp, clumps like cotton wool may form and turn into brown or dark red sponge-like growths that often are greater than 30 cm in diameter. On the surface of the sponge-like mass, seeds are produced which spread the disease to the other parts of the building.

Blue stain is one of a few relatively harmless fungi which cause stains on wood. It appears as a light blue discolouration, usually in lighter coloured wood such as Wawa, in sapwood and sometimes in unseasoned timber. The strength of the timber is not affected by this stain.

NOTES:

PREVENTION OF FUNGAL DISEASES

It is always better to prevent disease in healthy wood than to wait until disease is present and then try to treat it. Infection of timber can be prevented by following some simple precautions:

- It is important to ensure that fungi cannot find the conditions that they need to live, namely warmth, dampness and poor air circulation. Therefore the places where timber is built-in should always be dry and well ventilated.
- Use only healthy, well seasoned timber.
- Workpieces should be designed and constructed so that water drains quickly away from the wood.
- Use paint or preservatives on the wood if possible.
- Wood should not be in direct contact with concrete or masonry. If that cannot be avoided, the wood must be treated with a preservative. It is best to also use tarred paper to separate the wood from the concrete or plaster.
- Good storage, especially of unseasoned timber, helps to prevent attack by fungi (see Timber Piling).

REMEDIES FOR FUNGAL ATTACKS

Dry rot is very difficult and expensive to get rid of, once the timber is infected.

- First find if the timber is still strong enough to serve its purpose, if not replace it.
- Cut off and burn the affected parts.
- Find the cause of the dampness and provide good ventilation to the area (for example, repair defective roofs). Repair the damaged areas with new timber.
- Apply preservative to the timber.
- For blue stain, simply remedy the damp conditions, and provide good ventilation to the wood.

NOTES:

INSECT ATTACK

Most damage to wood by insects is caused by members of the beetle family and by termites.

BEETLES

The eggs of the beetles are laid in cracks on the wood surface and they develop into grubs or larvae. The larvae damage the wood by making small holes in the surface and then digging tunnels into it. They chew the wood and convert it into powder. Small piles of wood powder are pushed out of the holes and these are the sign that the wood is infested. When the larvae have eaten their way through the wood, they will leave it and fly away as fully developed beetles.

PREVENTION OF BEETLE ATTACK

It is quite difficult to prevent attack by beetles, especially in Wawa wood, since most beetles can fly. Some simple precautions can help.

- Paint, varnish, wax or wood preservative should be applied on all surfaces. The smell often keeps insects away. Chemicals are available which protect the wood; you should always follow the manufacturer's directions in using these.

- Beetles usually attack the sapwood first, because it is softer than the heartwood. This is why the sapwood should always be cut off.

- If whole logs, poles or sticks are to be stored, remove the bark first. Insects quickly multiply in wood from which the bark is not removed.

REMEDY FOR BEETLE ATTACK

When you suspect an attack, immediately inspect the wood for beetles. Check whether the wood is still strong enough to serve its function. There are chemicals available to kill the insects. Use them carefully and follow the manufacturer's instructions, as most of them are poisonous.

TERMITES

The greatest damage to wood here in Ghana is done by termites. They build their tunnels from the soil into the timber, leaving the surface of the timber untouched, which makes it very difficult to detect an attack in the early stages.

PREVENTION OF TERMITE ATTACK

The best way to prevent termites from attacking wood is to make sure that they cannot reach the wood.

- The wood should never be in contact with the soil, it should always rest on concrete etc.
- All sapwood should be cut off, because that will be attacked first.
- Protect the wood with wood preservatives.
- Do not use Wawa for construction wood, because it is very likely to be attacked.
- When wood must be in contact with the ground, for example with fence posts, it can be partly protected by scorching it over a fire, or by adding ashes around it when you set it in the hole. Termites do not like scorched wood or ashes.
- When wood is stored for seasoning etc., the ground under the stack should be covered with ashes.

REMEDIES FOR TERMITE ATTACK

- Destroy the path of the termites from the soil to the wood.
- Check if the wood is still strong enough for its function.
- Apply a wood preservative or a chemical to kill the termites.

PRESERVATION AND PROTECTION OF TIMBER AND MASONRY

It is important for the Rural Builder to protect timber and masonry in some way, to make them last longer. There are two basic types of protection we use:

- timber preservatives
- protective finishes

Timber preservatives are used only for wood, and they penetrate into the wood. The deeper they penetrate the wood, the better they work.

Protective finishes are used for both timber and masonry. They work by covering the surface with a protective "skin". Protective finishes are discussed on pages 200 and 201.

NOTES:

TIMBER PRESERVATIVES

Wood used in construction is often destroyed by fungal diseases or insects, especially termites. It is very important for a builder to find ways to protect wood from these dangers.

Some methods of protection have already been discussed in the sections on fungal and insect attack. There we mentioned the uses of wood preservatives. Wood is food for fungi and insects. This food can be poisoned for them by wood preservatives. The wood absorbs these preservatives easily and the fungi and insects that try to eat the treated wood will die.

There are two classes of preservatives: waterborne preservatives and oil preservatives. The type we choose will be determined by the intended use of the wood and by what further surface treatment (painting, etc.) will be done.

- WATERBORNE PRESERVATIVES: These are usually available as powders which are dissolved in water and applied to the wood. Since water is the base the preservative can also be washed out again by water. This can happen if rain should reach the wood. For this reason use waterborne preservatives only under dry conditions and not for outside work where the rain can wash them out.

  Kinds of waterborne preservatives:

  - Aldrex 40 (mix 1 part of Aldrex 40 with 40 parts of water; use $1\frac{1}{2}$ Ideal milk tins of the powder in a No. 28 bucket of water, which will give the correct proportions)

  - Any other chemical preservative which is mixed with water.

- OIL PRESERVATIVES: These have an oil base. They not only kill insects and fungi, but also keep water from penetrating the wood. They do not wash out with water, so they are useful for outside work.

  The most common oil preservatives are:

  - solignum

  - creosote

  - used engine oil

  - Aldrex 40 mixed with engine oil or kerosene; 1 part Aldrex to 40 parts oil or kerosene.

- NOTE: While it is possible to apply an oil preservative over wood that has already been treated with a waterborne type, paint or waterborne preservatives cannot be applied over oil preservatives because they cannot penetrate.

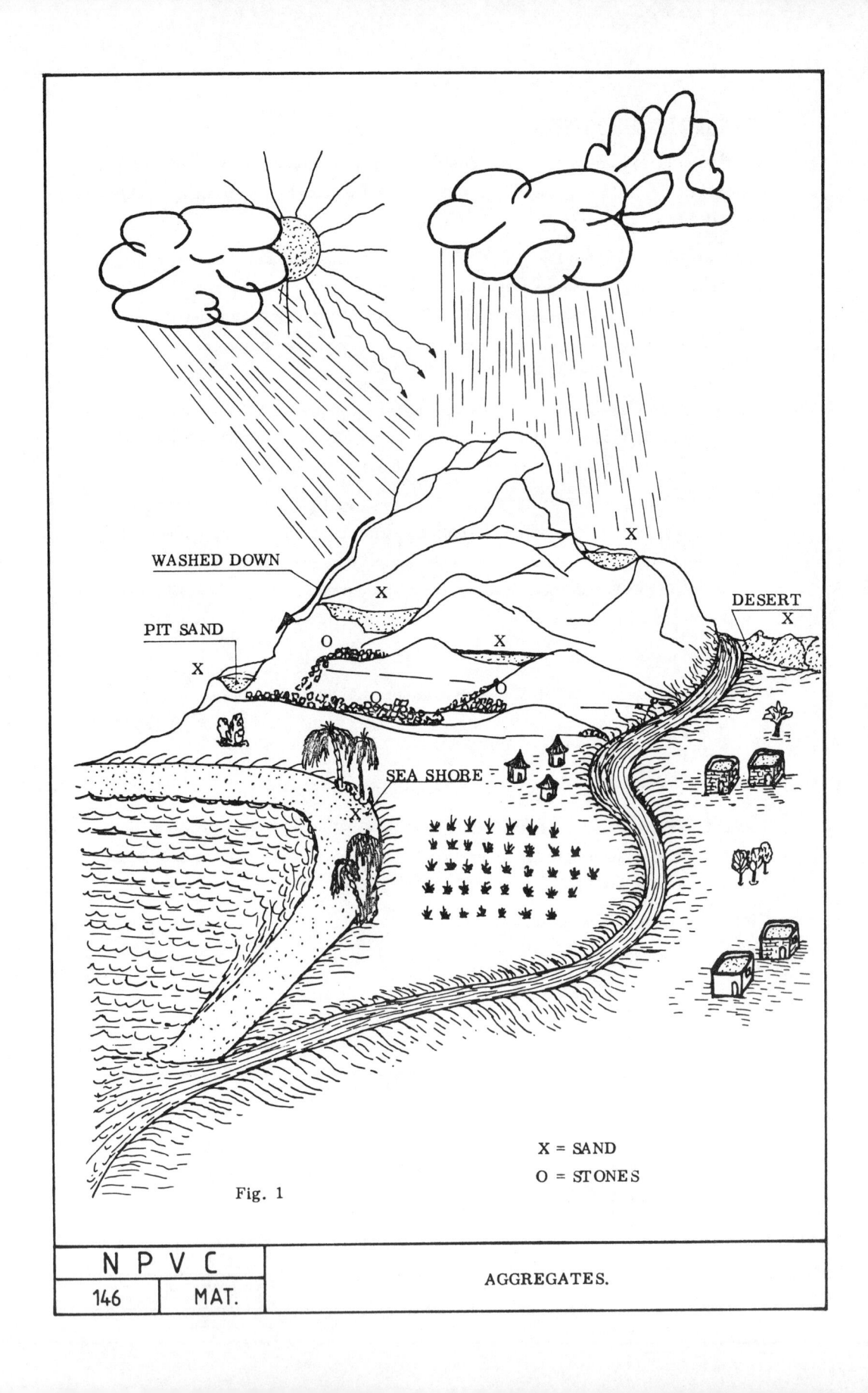

Fig. 1

X = SAND
O = STONES

NPVC		AGGREGATES.
146	MAT.	

AGGREGATES

"Aggregate" is the term used for the mixture of different sized stones that form the body of mortars and concrete. Ideally the stone should be graded so that the smaller sizes of stone fit exactly into the spaces between the larger ones and no gaps or holes are left in the mass of mortar or concrete.

SAND

Sand is a mass of finely crushed rock. It is either crushed naturally as seen on the sea shore, in river beds, or in deserts (Fig. 1); or it is artificially produced in crusher plants near rock quarries (where rock is dug out of the earth).

Sand is classified according to the shape of its particles (which differs depending on where the sand came from originally). It is also graded according to the size of its grains (the individual particles of sand).

GRAVEL

"Gravel" is the term commonly used for the larger sized stones of the aggregate. Originally, gravel meant an "all-in-one" aggregate, a mixture of sand and stones of all sizes which can sometimes be found all together in a natural deposit. The individual particles are rounded by the natural action of water and weather.

BROKEN STONES

These are the largest stones of the aggregate, they make up the bulk of concrete. They are found either in natural deposits or scattered on the ground surface; or they are artificially produced in crusher plants. The Rural Builder often must break up large stones with hammers, to make them a convenient size (see Tools, page 14).

- NOTE: These aggregates are the most common ones used for building in the Northern and Upper Regions of Ghana. Of course there are many other types of aggregates (chips, pebbles, rubble etc.), but as far as the Rural Builder is concerned they are of little importance.

NOTES:

QUALITY AND PROPERTIES OF AGGREGATES

Good mortar and concrete can never be made with poor materials. The cement, sand and stones must all be good quality and the correct types. Sand and stone (the fine and coarse aggregate) together make up more than four-fifths of the concrete mass, so there can be no doubt about their importance. It is not safe to take for granted that every load of sand or gravel brought to the site will be up to standard. Remember that aggregates are either dug from a pit or river bed or they are quarried, and although they may look the same there is a possibility of variation in the quality of different loads.

Particle sizes, the shape and texture of the particles and their surface areas are all important factors in the strength and durability of the concrete or mortar.

- GRADING: A graded aggregate is one that is made up of stones or particles of different sizes, ranging from large to very small. It sometimes happens that a load of sand will have too many coarse particles to make a good mortar, while another load will have too many fine particles. Depending on the job to be done, you might have to mix the two sands together in different proportions to get a suitable aggregate. If the sand contains too many bigger particles it may be necessary to sift these out before using the sand to make mortar, but it could work well for concrete.

The idea is to come up with a "well graded" aggregate; which means that the smaller grains will fit in between the larger ones, leaving only small spaces to be filled with the cement paste. The result will be a good workable mix of adequate strength, using a minimum amount of cement (see sketches below).

FINE UNIFORM

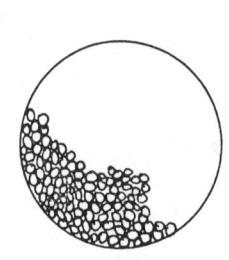

COARSE UNIFORM

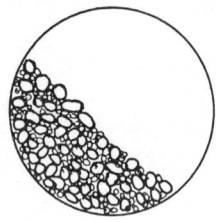

WELL GRADED

MIXED WITH CEMENT PASTE

AGGREGATES.

- CLASSIFICATION OF AGGREGATES: For making concrete and mortar, the Rural Builder has two types of aggregate: the fine one which is sand; and the coarse one which is broken stones. Both aggregates are classified according to their grain size and are each divided into two main groups:

Fine sand -- from 0 - 1 mm	Fine broken stones -- from 5 - 25 mm
Coarse sand -- from 1 - 5 mm	Coarse broken stones -- from 25 - 50 mm

Another classification is made according to the shape and texture of the single particles. Some sands and stones have particles which are rounded, with relatively smooth surfaces. This sort of aggregate is found mainly in river beds, along the shores of lakes and coasts, and in deserts. This weather- and water-worn sand is called "river sand" or, because of its properties and workability, "soft sand" (Fig. 1).

The other type of sand has a fairly rough surface and it is found mainly in deposits close to hills and mountains. Artificially made sand made from crushed rock also comes under this classification. It is known as "pit sand" or else "sharp sand" (Fig. 2).

Whether the sand is soft river sand, or sharp pit sand; it will have various grain sizes and is classified as fine or coarse, as in the table above.

- NOTE: In the Northern and Upper Regions of Ghana, most sand is dug from river beds. This does not necessarily mean that the sand will be "soft". Often it is a sharp sand or between soft and sharp, because the particles don't get exposed long enough to weather and water to become rounded and smooth.

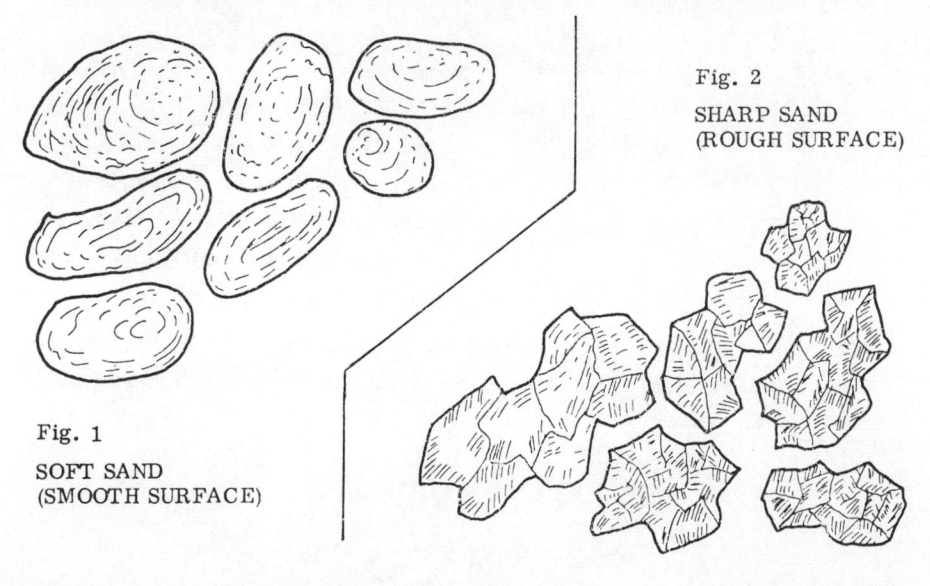

Fig. 1
SOFT SAND
(SMOOTH SURFACE)

Fig. 2
SHARP SAND
(ROUGH SURFACE)

AGGREGATES.

N P V C
MAT. 149

An important factor in the quality of an aggregate is its cleanliness. Clay, mud, or fine dust (known as silt) in the aggregate will weaken the concrete or mortar; while any rotting vegetable matter (organic impurities) like leaves, grass or roots may interfere with the setting of the cement.

- THE HAND TEST FOR SAND: As a first test for cleanliness, simply pick up a little sand and rub it between your hands. If your palms stay clean, the sand is clean enough. If not, the sand may contain too much silt.

- THE SILT TEST FOR SAND: You yourself can carry out a simple test to get an idea of the amount of silt in a natural sand (though not in an artificially crushed rock sand).

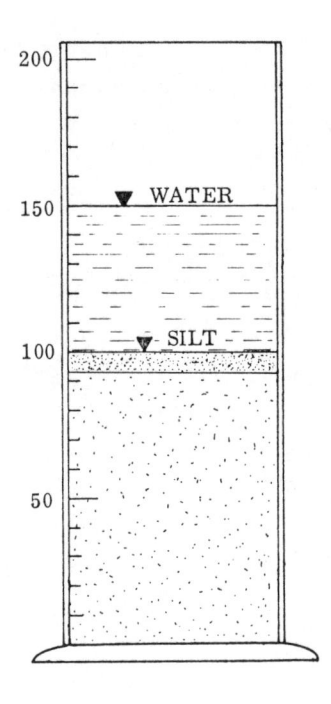

To test accurately you should have a measuring cylinder which is marked in millilitres, shown as "ml", usually up to 200 ml.

First make a salt-water solution by putting one teaspoon of salt into 1/2 litre of water. Fill this solution into the measuring cylinder, up to the 50 ml mark.

Next pour in the sand until the level of the sand is up to the 100 ml mark. Add more salt water until the water reaches the 150 ml mark, cover the cylinder and shake it well.

Stand the cylinder on a level surface and tap it gently until the top surface of the sand is level. Leave it to settle for 3 hours and then measure the height of the silt layer on top of the sand. This should be no more than about 6 ml, or about 6% of the total amount of sand.

- NOTE: If the sand contains more than 6% silt, you would have to use more cement and the concrete would shrink more during the hardening process, causing cracks in the product.

N P V C		AGGREGATES.
150	MAT.	

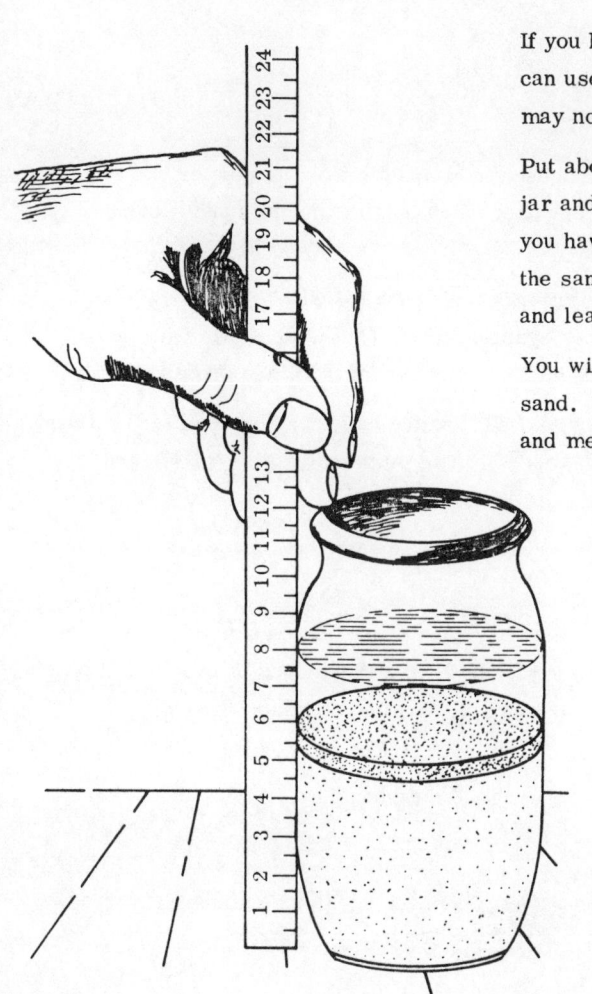

If you have no measuring cylinder, you can use a 0,5 kg jam jar, though this may not be quite accurate.

Put about 5 cm of sand loosely into the jar and pour some salt water on it until you have about 2,5 cm of water above the sand. Now cover and shake the jar, and leave it to stand for about 3 hours.

You will see a layer of silt on top of the sand. Measure the depth of the layer, and measure the sand below it. There should be no more than about 3 mm of silt, or about 6% of the amount of sand.

- ORGANIC IMPURITIES: The Rural Builder can carry out a test for organic impurities using a glass jar. Put sand into the jar and fill up the rest of the jar with water. Cover and shake the jar and leave it standing for some minutes. If the water above the sand is brown or very dirty, the sand contains organic impurities and cannot be used.

Better sand can be found by simply removing the top layer of sand, about 5 cm deep, before taking sand from a dry river bed. This top layer consists mainly of excessive silt as well as organic impurities such as vegetable matter and cow dung. None of this is wanted, because it would cause problems with the concrete or mortar.

- REMEMBER: Wherever your sand comes from, it must be clean and suitably graded. If you use dirty sand, you may find that it mixes very nicely, but you will find problems before the job is finished. The impurities in it may affect the rate of setting and hardening of the concrete or mortar, and decrease the final strength of the work. The fine appearance of the just finished work may be spoiled by cracking and flaking as it dries.

AGGREGATES.	N P V C	
	MAT.	151

BINDING MATERIALS

LIME

Lime is a very fine white powder, used in mixes for mortar, plaster and render. It is made from limestone or chalk which is burnt in a kiln and becomes quicklime.

The quicklime is usually passed through a machine called a hydrator, where it combines with water and becomes hydrated lime. This is dried, crushed to a fine powder, then bagged and sold. Below is a diagram of the whole process.

Sometimes the lime is sold as quicklime, and the builder adds the water to it himself. This process is called "slaking" the lime or "running it to putty", and it is not described here.

Slaked lime and hydrated lime are chemically the same, but slaked lime has more water in it.

Hydraulic lime is made from limestone or chalk containing clay. It hardens when combined with water; and it also hardens well in damp places or even under water. It is stronger than other limes, although weaker than Portland cement.

Non-hydraulic lime comes from the purest limestones and chalks. It hardens by drying out and then slowly combining with the carbon dioxide in the air.

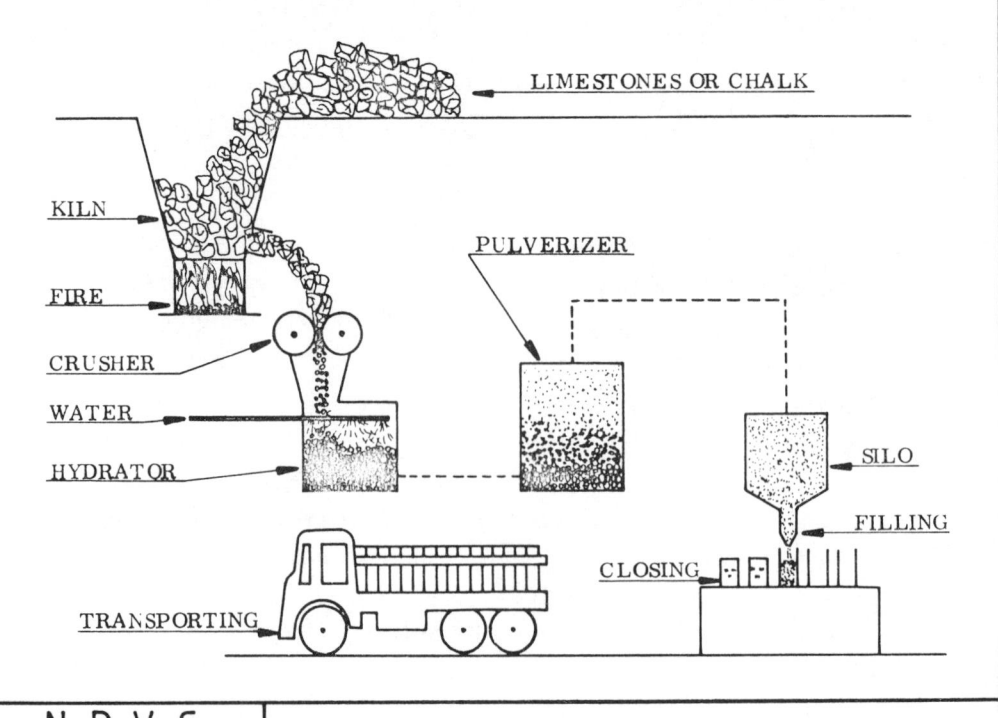

BINDING MATERIALS.
NPVC 152 MAT.

PORTLAND CEMENT

Portland cement is a fine grey powder. Among the various kinds of cements, it is the most commonly used as binding material. It is made of a mixture of chalk or limestone and clay.

The following description of the manufacturing process is illustrated below.

The limestone or chalk and the clay, in appropriate proportions, are fed into a "wet grinding mill" and reduced to a creamy substance known as slurry. The slurry is pumped to a large cylindrical "kiln" which is about 90 m long and 3 m in diameter. The slurry enters the kiln at its upper end while pulverized (crushed) coal, gas or other fuel is blown in at the other end. The temperature inside the kiln at the lower end is very intense, approximately 1500 degrees C; gradually decreasing towards the top end. So the slurry as it moves down the kiln is first dried, then heated, and then finally burnt. It leaves the kiln in the form of very hard "clinkers" shaped like small balls and of a dark brown to black in color. The clinkers are ground up to an extremely fine grey powder, which is the cement. The cement is packed in paper bags of 50 kg capacity.

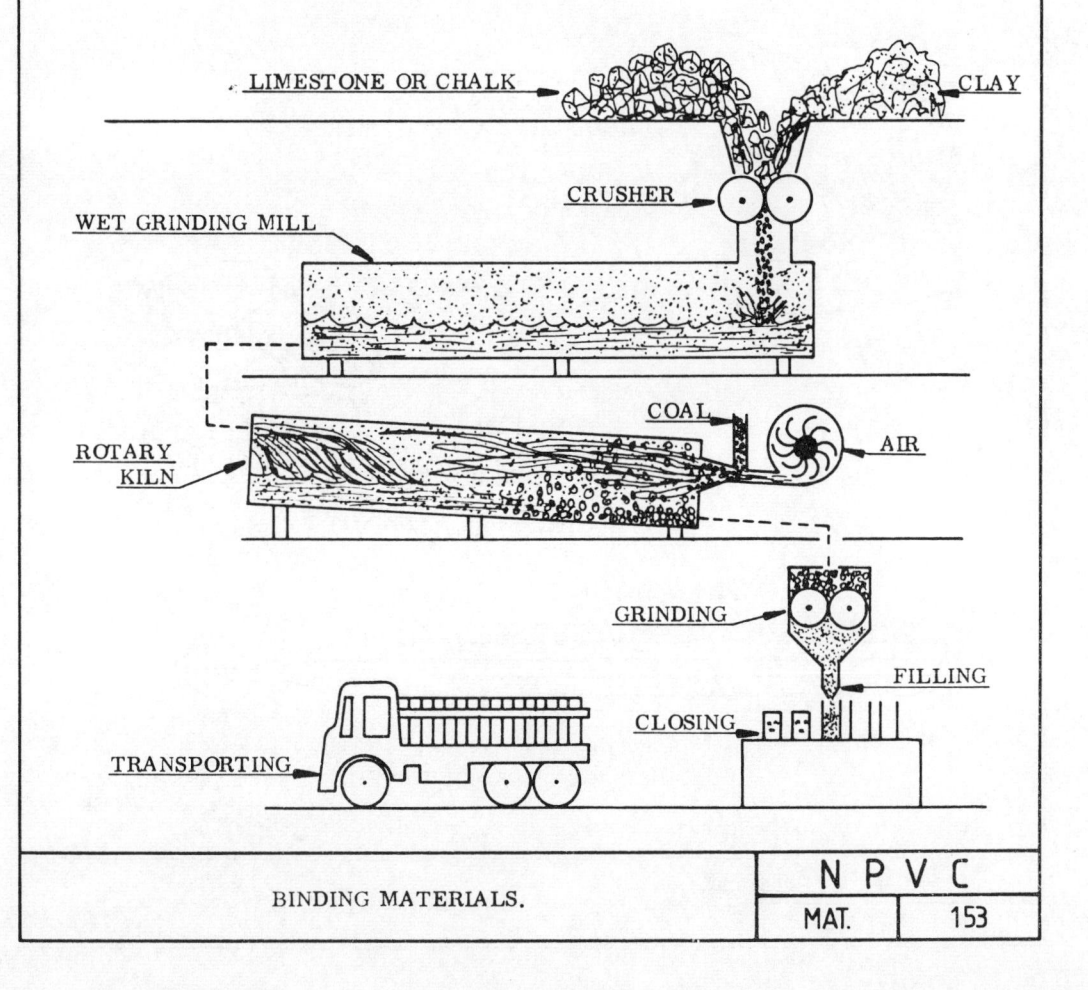

BINDING MATERIALS.

NPVC
MAT. 153

HYDRAULIC CEMENTS HARDEN

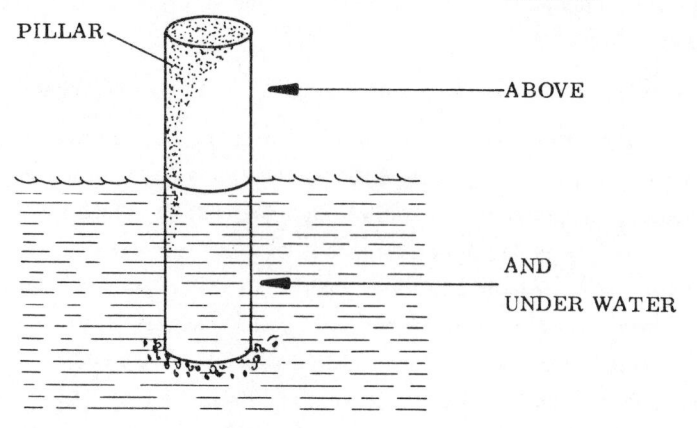

PILLAR

← ABOVE

← AND UNDER WATER

Fig. 1

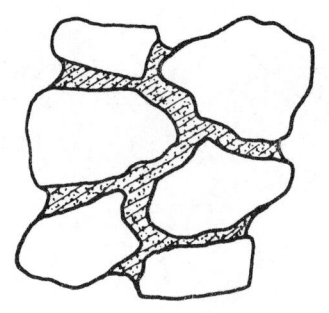

Fig. 2a
HARDENED CEMENT PASTE BINDS STONES TOGETHER AND FORMS A SOLID MASS

Fig. 2b
BITUMEN OR TAR ALSO BINDS STONES TOGETHER BUT REMAINS A BIT SOFT (WATCH FOR EXAMPLE TARRED ROADS IN THE HEAT)

← GLUE

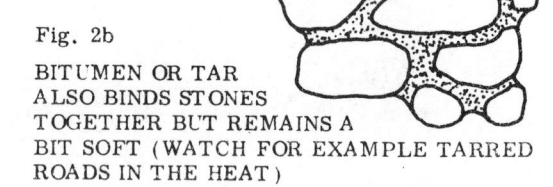

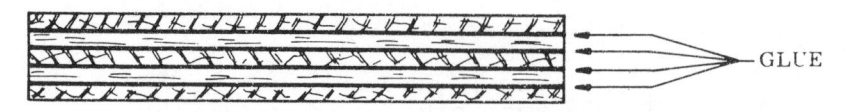

Fig. 2c   GLUE HOLDS WOOD, PLASTIC, ETC. TOGETHER TO MAKE JOINTS OR SOLID PIECES, FOR EXAMPLE PLYWOOD

N P V C		
154	MAT.	BINDING MATERIALS.

HISTORY OF CEMENT

Some sort of binding substance has been used since ancient times to hold together the stones, bricks etc. used in building. The earliest building cement was probably clay or ordinary mud. The Romans were master builders in brick and stone, and a large part of their success was because of their discovery of a cement that was made by mixing a volcanic ash with burned lime. The Romans also made pure lime mortars and gypsum plasters. These materials were the only building cements until modern times.

The modern era of building cements began about 1760, when an English engineer discovered the most suitable composition for hydraulic cements. These are cements which will harden even under water (Fig. 1). A few years later, in 1824, another Englishman invented Portland cement. He named it because of its similarity in appearance to a natural stone from Portland in England.

- DEFINITION: A cement is any material which attaches or unites two surfaces, or serves to combine particles into a whole.

- TYPES OF CEMENTS (Fig. 2):

    a - Building cements (eg. Portland cement and lime)
    b - Bituminous cements (eg. tar and asphalt)
    c - Adhesives (eg. animal glues and synthetic resins)

NOTES:

BINDING MATERIALS.

N P V C

MAT. 155

Fig. 1
HOW TO STACK BAGS OF CEMENT IN A STORE

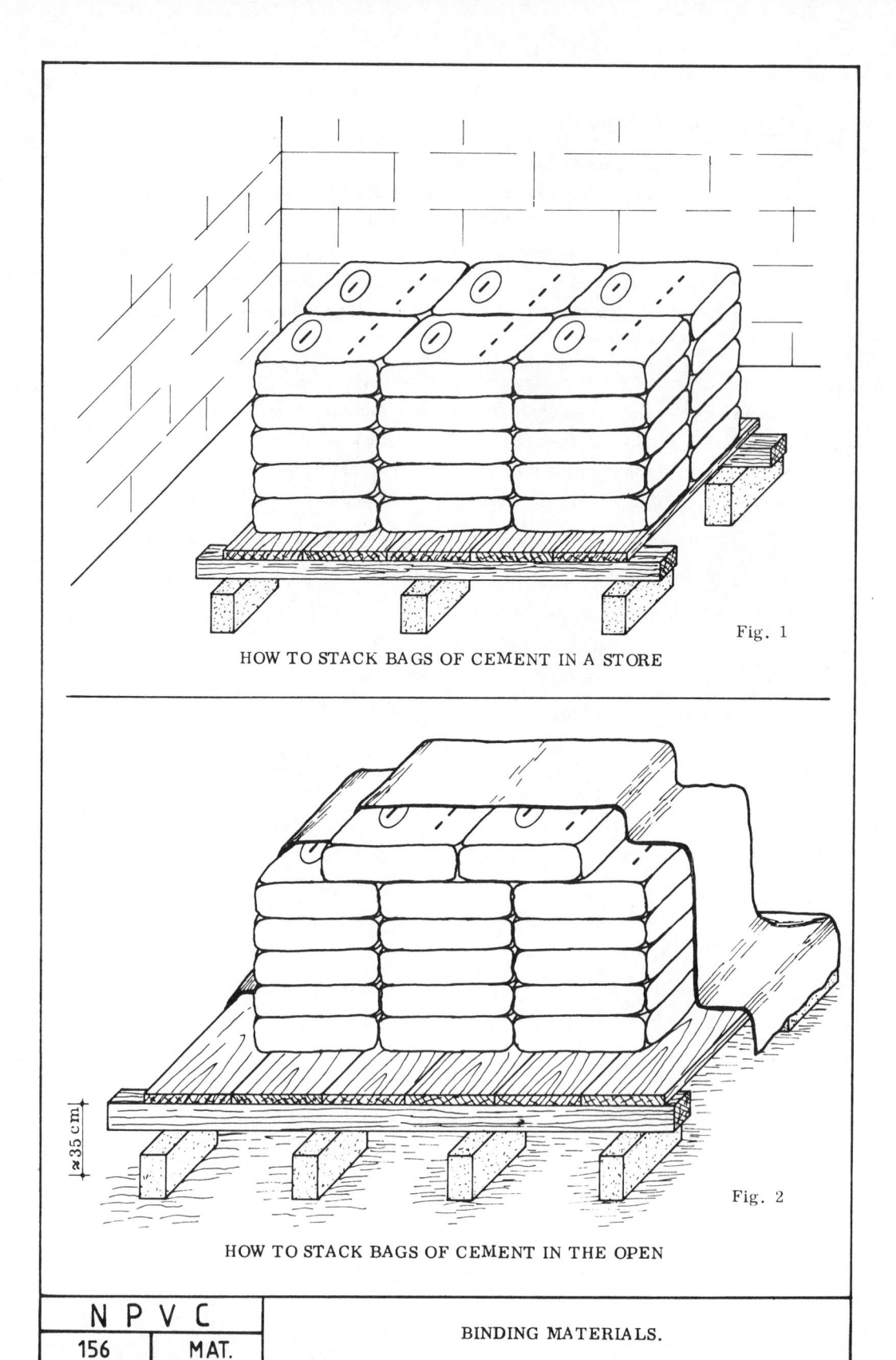

≈35 cm

Fig. 2
HOW TO STACK BAGS OF CEMENT IN THE OPEN

N P V C		
156	MAT.	BINDING MATERIALS.

STORING BINDING MATERIALS

The quality of mortar and concrete depends on so many factors, but one of the most important of these is the cement. Cement must be stored properly, to prevent it from setting (hardening) before it is used. If the cement gets damp, it will become unusable. Everyone knows that cement should be kept dry, but they don't always realize that contact with damp air can do as much harm as direct contact with water. On all jobs where bagged cement is used, there should be a shed or room to store it.

- STORING IN A SHED: Make sure that the shed or room is water-tight and has a sound, dry floor. If the floor is not dry, make a platform out of boards set on blocks and timber, to raise the bags off the ground (Fig. 1). Stack the bags closely together to keep out air, and away from the walls so that they are not in contact with any dampness on the walls. In very large sheds it is better to cover the bags with plastic sheeting to keep out damp air, especially during the rainy season.

Check the bags from time to time for termites: these may damage the bags and with them the cement. Check also that the roof doesn't leak and that the walls are waterproof.

- STORING IN THE OPEN: On some jobs, bags of cement may have to be stored in the open, with no more protection than a dry base and a covering of tarpaulins or plastic sheets. The sheeting must be properly overlapped to keep out the rain; and the top sheet should lay over all the ones below like a roof, so that the rain can run straight off without getting into the tarpaulin "tent" and wetting the cement.

Even if the cement is to be stored in the open for only an hour or so, there must be a dry platform raised about 35 cm above the ground for the bags to lie upon (Fig. 2).

Whether the cement is stored indoors or out, arrange the bags so that the first batch brought in can be the first ones used, and the old bags don't get left at the bottom of the stack and never used. Prevent accidents by keeping the piles to a height of about 1,20 m, and never stack them more than 10 bags high.

- REMEMBER: The important thing with cement is to always KEEP IT DRY! Cement starts to set about 30 minutes after mixing or coming into contact with water or moisture.

MORTAR

Mortar consists of the body or aggregate, which is fine sand; and the binding material, which is cement mixed thoroughly with water.

Mortar is used to bed blocks as well as for plastering. A good mortar should be easy to use and should harden fast enough that it does not cause delays in the construction. It must be strong enough, long lasting and weatherproof.

TYPES OF MORTAR

The best mortar for a particular job is not necessarily the strongest one. Other properties like workability, plasticity or faster hardening can be more important, though the strength of the mortar must of course be sufficient for the job.

Mortar should neither be much stronger or much weaker than the blocks with which it is used.

- CEMENT MORTAR: This sets quickly and develops great strength. It is used in proportions of one part cement to three parts sand (1:3), which makes quite a strong and workable mix; down to a 1:12 mix, a lean mix which will be rather harsh and difficult to use.

- LIME MORTAR: This is usually very workable and does not easily lose water to the blocks, but it is weaker than cement mortar and hardens slower. Lime mortars are nowadays largely replaced by cement mortars or combinations of lime and cement.

- CEMENT-LIME MORTAR: This combines the properties of cement and lime to give a workable and strong mortar. The cement makes the mortar stronger, denser, and faster setting; while the lime makes the mortar workable and reduces the shrinkage during drying, because it retains the water better.

In some areas, lime is not always available, so in this book we will concentrate on the use of cement mortars.

NOTES:

SELECTING THE RIGHT KIND OF SAND

Sand for plaster, mortar and renderings must always be chosen with care. The sand used to make mortar for blocklaying should be well graded, sharp and must not be too fine if a strong mortar is needed (eg. for footings).

The more fine particles the sand contains, the better its workability in the mix, but more cement paste will be needed to cover the surfaces of the particles. This means that in order to improve the workability while maintaining the same strength, more cement must be added which results in higher costs.

The Rural Builder is always faced with this problem and it takes a lot of experience to be able to find a good compromise.

If the sand is found to be too sharp so that it makes a mortar with poor workability, we suggest replacing about 1/3 of it with fine soft sand; but don't replace more than about 1/2 unless you add more cement.

We can do this because the common mix proportion of our mortar is 1:6, while the sandcrete blocks are mixed in a proportion of 1:8 (cement:sand). When the fine sand is added, the strength of the mortar is reduced to about the same as the strength of the blocks, which is acceptable.

The information above is meant to show the problems concerning the selection of the right sort of sand for a particular job. This book will address these problems repeatedly as we introduce the different kinds of blocks as well as the different jobs.

- REMEMBER: A good mortar should:
  a - be easy to work with
  b - harden fast enough not to delay the construction
  c - stick well to the blocks
  d - be long lasting and weatherproof
  e - if possible, be as strong as the blocks.

NOTES:

STAGE 1
FILL FLUSH

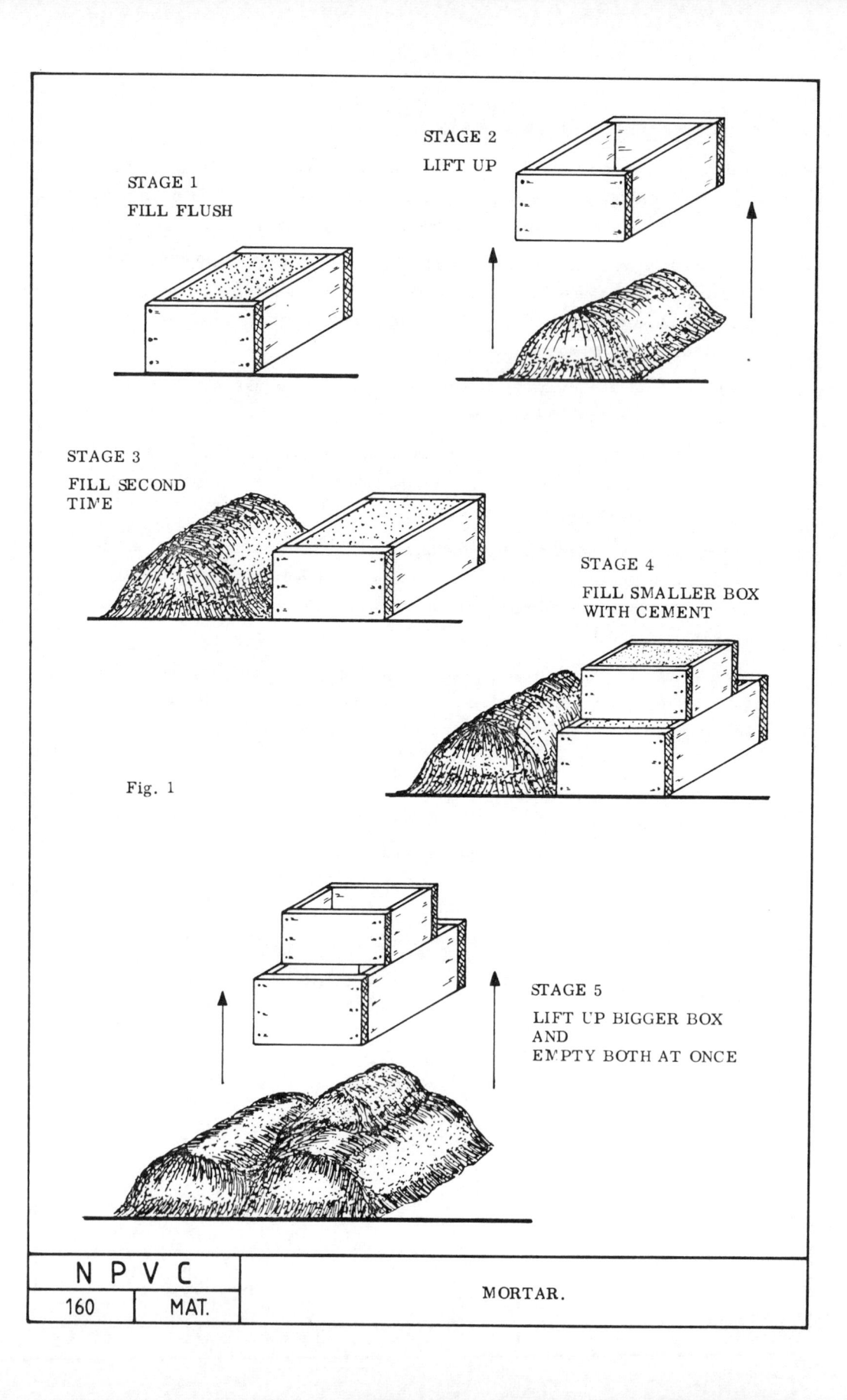

STAGE 2
LIFT UP

STAGE 3
FILL SECOND TIME

STAGE 4
FILL SMALLER BOX WITH CEMENT

Fig. 1

STAGE 5
LIFT UP BIGGER BOX AND EMPTY BOTH AT ONCE

N	P	V	C
160	MAT.		

MORTAR.

BATCHING

By the term "batching" we mean that we measure the proportions of the various ingredients of a mix. We already know that the ingredients for a mortar should be mixed in certain proportions. To help us to obtain the correct amounts we can make boxes with the appropriate sizes; this method is known as "batching by volume".

Since a common mix proportion is 1:6 and some special jobs require a mix of 1:3, two batching boxes are made with corresponding measurements. The smaller one for cement measures 15 cm high, 30 cm wide and 38 cm long (inside measurements) giving a volume of 17,1 litres, which is about half a bag of cement or one slightly heaped headpanful (one bag of cement contains 34,1 litres or two slightly heaped headpans).

For the sand, a bigger box is made which must hold exactly three times as much as the smaller one. Therefore its inside measurements are: 30 cm high, 30 cm wide, and 57 cm long. This gives a volume of 51,3 litres, which is 3 x 17,1.

Both boxes are bottomless; they are only frames in order to make it easier to work.

- PROCEDURE: Before you start batching and mixing, it is advisable to make a mixing platform out of lean concrete for all future mortar and concrete work. This provides a firm and clean place so that your mix will not become dirty, no cement paste will be lost, and shovelling will be easier.

For small jobs, the mixing platform can measure approximately 1,5 m by 2 m and may be 5 to 7 cm thick.

For a 1:6 mix, place the bigger box on the platform and fill it with sand flush to the top edge. Then lift it up and set it down again next to the first pile and refill it in the same way (Fig. 1, stages 1 to 3).

Now put the small box on top of the sand-filled big one, and fill it with cement. It must be filled flush to the top edge in order to get the right proportions (Fig. 1, stage 4).

Now lift up the lower box, taking the smaller one with it and emptying both at once. The result is a heap of sand (6 parts) covered with cement (1 part), ready to be mixed (Fig. 1, stage 5).

Fig. 1   SAND AND CEMENT

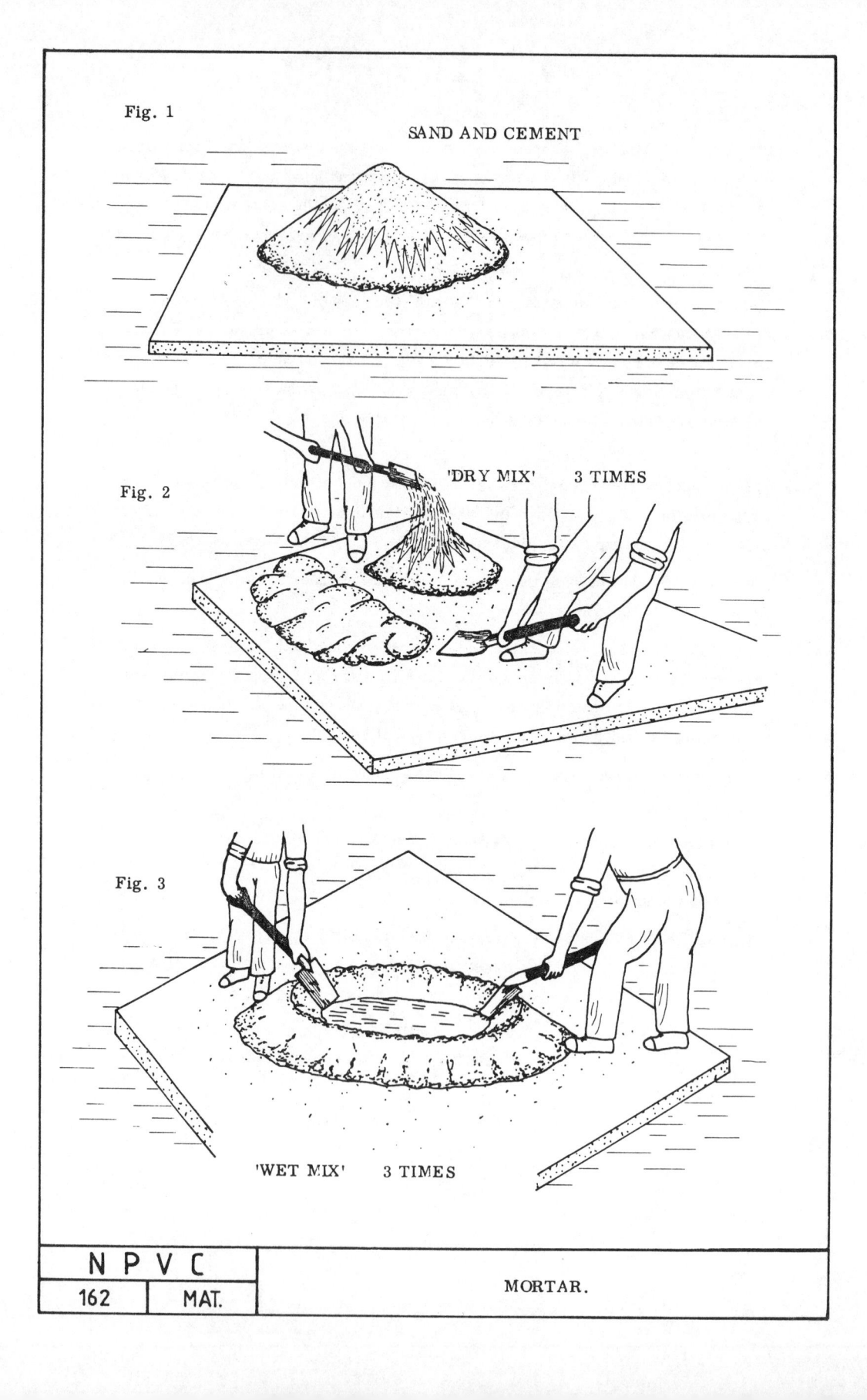

Fig. 2   'DRY MIX'   3 TIMES

Fig. 3   'WET MIX'   3 TIMES

N P V C	
162	MAT.

MORTAR.

MIXING THE MORTAR

Mixing is one of the most important stages in the process of making mortar because the workability and strength of mortar depend so much on the way it is mixed and on the amount of water added to the mix.

- WHAT WATER DOES: Water in the mix does two things: it makes the mortar workable and it combines chemically with the cement to cause hardening. However, only about half the water is required for the chemical reaction and the rest will remain or evaporate slowly as the mortar hardens, leaving small holes or "voids" in the cement.

  Obviously, the more water there is in the mix, the greater will be the number of voids and the weaker the mortar.

  For your 1:6 mixture, a maximum of 15 litres of water should be added (almost one headpanful); never any more than this even if the mix appears to be too dry, as sometimes happens in the first stage of mixing.

- THREE TIMES DRY: The sand and cement is measured on one end of the mixing platform. With two men facing each other across the pile and working their shovels together, turn the whole heap over once to form a pile at the other end of the slab (Figs. 1 & 2). This turning must be repeated twice and results in a so-called "dry mix".

  The correct method for turning over is to slide the shovel along the top of the platform, pick up a load and spill the load over the top of the new pile. The main point is that each shovelful runs evenly down the sides of the cone. This is the best and easiest way of mixing dry mortar and all other motions should be eliminated. When the dry mix is a uniform colour throughout, it is considered to be well mixed.

- THREE TIMES WET: Form the heap of dry mix into a crater or pool, with the sides drawn out towards the edges of the mixing platform. There should be no mixture left in the centre of the pool.

  Now gently pour about 3/4 of the total required water into the crater. Turn the shovel over and with the edge scraping along the platform, push some of the dry mix into the pool in such a way that it spreads out, without separating the sand and cement. Handle the shovel carefully so that no water can escape by breaking through the ring (Fig. 3).

  When all of the dry mix has been heaped up in the centre of the platform, it should have taken up all the free water and have a rather stiff consistency (earth-moist).

Now make a second pool, add the remaining water and repeat the rest of the mixing procedure. This will result in a mortar of a plastic consistency. To make sure the mixing is thoroughly done, turn the mortar over a third time.

- CONSISTENCY TEST: You can carry out a simple test to get a rough idea whether the consistency of the mortar is correct (this means the water content).

Fill a headpan with mortar and smooth the surface (Fig. 1). With the blade of your trowel, make a straight cut clear through the mortar to the bottom of the headpan (Fig. 2). Now push the trowel flat under the mortar along the bottom of the headpan, so that the cut in the mortar centres the length of the trowel.

When the trowel is lifted up 2 or 3 cm, the gap in the mortar must open into an oval shape along the outer edge but remain closed along the bottom (Fig. 3).

- NOTE: If you want to improve the workability of the mortar by adding water, remember that this will also decrease its strength. You are therefore strongly advised to add both cement and water in equal quantities (for example 1/2 bucket of water plus 1/2 bucket of cement).

Fig. 1

FILL HEADPAN HALF FULL AND SMOOTH THE SURFACE

Fig. 2

MAKE A CUT WITH THE TROWEL

Fig. 3

MORTAR.

- REMEMBER : Ready mixed mortar starts setting after only 30 minutes! Never prepare more mortar than you can use within this time.

It is certainly better to mix smaller amounts more often than to allow mortar to spoil; or to do the work very quickly (and sloppily) in order to get rid of the mortar.

Always cover freshly mixed mortar with empty cement bags to keep it from drying out.

NOTES:

CONCRETE

To concrete something means to form it into a mass, or to solidify it.

As far as building is concerned, the term concrete means an artificial stone made by mixing sand, stone, Portland cement and water. This mixture, cast into a form of the desired shape and size, hardens into a stone-like mass: the concrete.

There are basically three materials we start with to make concrete:

- The aggregate, which is made up of the fine and coarse aggregates together, ie. the sand and broken stones. The aggregate makes up the main mass of the concrete; its function is mostly just to add bulk.

- The water.

- The binding material, which is usually Portland cement.

When the three materials are mixed together, the cement and water combine chemically to make a cement paste, which surrounds the particles of the aggregate and holds them together.

NOTES:

CEMENT PASTE

The cement paste component of concrete is what causes it to harden, the aggregate simply remains passive (inactive).

Thus the cement paste must completely cover the surface of every single particle of the aggregate. This means that each stone, no matter whether tiny or big, must be covered all over by a thin layer of cement paste.

This is achieved by mixing all three components very thoroughly and in the correct proportions (see Batching, and Mixing The Mortar, pages 160 to 163).

The cement paste fills up all the spaces between the particles of the aggregate and bonds them firmly together as it hardens.

The hardening process requires a certain amount of water; how much depends on how much cement is added to the mix. The correct proportions can be found in the Tables of Figures, page 234.

After it is set, the hardened cement paste cannot be dissolved again (except by the use of certain acids).

An undesirable further reaction of the cement paste is the drying shrinkage as it hardens. Because of the evaporation of the extra water, the volume of the concrete is gradually reduced. The concrete shrinks and develops cracks.

This reaction can be effectively reduced, if not prevented, by correct curing; as will be discussed later in this book.

Also to prevent cracking, large areas that are covered with concrete; such as floors, should be divided up into bays.

PROPERTIES OF CONCRETE

Concrete has many properties, but most of them are of little interest to the Rural Builder. Therefore this chapter deals only with the three most important properties:

a - Compression strength

b - Tensile strength

c - Protection against corrosion.

NOTES:

- **COMPRESSION STRENGTH:** It is commonly known that concrete becomes very hard and can withstand enormous pressures; a property which is called compression strength.

This compression strength depends mainly on the properties and quality of the cement paste and the aggregate.

- If the aggregate consists of a soft or weak material, the concrete will be weak also.
- If the aggregate is so dirty that there is no direct contact between the surface of the particles and the cement paste, the concrete will again be weak.

Provided that all the rules for producing a good concrete are observed, the strength of the concrete can be controlled by choosing the mix proportions. For example, a mix proportion of 1:10 is weaker than a 1:3 mix. This is because in a 1:10 mix the particles of aggregate are not completely coated with cement paste, but in the 1:3 mix they are fully embedded in it.

- If not enough water was added to the mix, the cement paste remains too dry and stiff and the concrete will be weak.
- If too much water was added, making the cement paste too thin, the concrete will again be weak.

Therefore the Rural Builder must always carefully follow the correct concrete recipe.

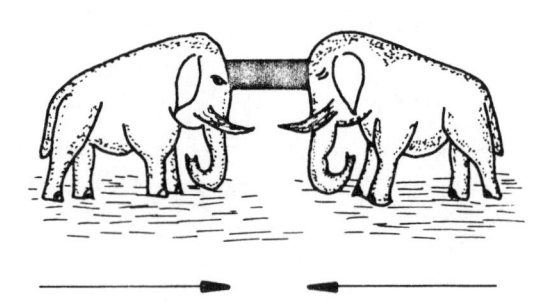

COMPRESSION STRENGTH

CONCRETE.

- TENSILE STRENGTH: The tensile strength of a material means its capability of being stretched to a certain extent without breaking.

Although concrete becomes very hard, its tensile strength is very limited. It is so low that in practice, the tensile strength of concrete is regarded as being non-existent. This is why sometimes concrete members of a structure must be reinforced by steel bars embedded in them.

Some types of wood, while they are softer and have a much lower compression strength than concrete, have a far higher tensile strength because of their fiber structure. The wood fibres act in a way like the reinforcement iron embedded in concrete.

Wood is a good building material because of its tensile strength. However, its flexibility makes it subject to bending under loads. Because of this problem, short-span constructions are chosen; or, among other possibilities, reinforced concrete can be used instead of wood.

TENSILE STRENGTH

NOTES:

CONCRETE.

NPVC
MAT. 169

- PROTECTION AGAINST CORROSION: Corrosion means a wearing away, a slow destruction caused by a reaction with air, water or chemicals.

Reinforcement iron which is left unprotected and exposed to air and humidity will eventually start to corrode on the surface and become rusty.

If this process is not halted in time, the rust goes into the bar and it becomes too weak to be used.

In order to maintain the strength of steel-reinforced concrete, the steel has to be protected from rust. This is partly done by the hardened cement paste and partly by structural means.

Ideally, the hardened cement paste hermetically seals the iron so that direct contact with air and humidity is cut off. Even slight rust stains on the iron cannot do any harm because the cement paste protects it against further corrosion.

The protection will not be enough however, unless the builder observes the following rules:

- The reinforcement bars must be completely covered by concrete which is well compacted and without voids.

- The concrete cover must be sufficiently thick, and without cracks.

- In most cases ordinary Portland cement is used and the mix proportion should be no less than 1:5 for reinforced concrete. (see Tables of Figures, page 234).

Apart from these, all the other rules for producing a good concrete must be observed.

- NOTE: Quality concrete is not a brand. It does not have a trademark on it to say "This is quality concrete". Sometimes the concrete does not even look different from poor concrete, but it is different. This depends not only on the mix proportion, but on the awareness and skill of the builder.

NOTES:

PART 4: RURAL BUILDING PRODUCTS

REINFORCEMENT STEEL

To reinforce a material means to add something to it, in order to make it stronger.

One of the strongest reinforcement materials available is steel or iron. In reinforced concrete, a concrete member is strengthened with steel bars or metal netting embedded in it.

TYPES OF REINFORCEMENT STEEL

There are various types of reinforcement steel; how they are used depends on the function, shape and dimensions of the reinforced concrete member as well as on the required strength.

Reinforcement steel is classified according to its shape and surface texture. The most common reinforcement is single round bars which can have either a smooth or a ribbed surface.

- CIRCULAR BARS: Round, smooth bars are called circular bars and are available in diameters ranging from 5 mm to 28 mm (Fig. 1, next page). The four sizes most often used in Rural Building have diameters of 6 mm (1/4"), 10 mm (3/8"), 12 mm (1/2") and 18 mm (3/4").

- RIBBED BARS: The round bars with a ribbed surface are called ribbed bars and are available in diameters ranging from 6 mm to 40 mm, if the bar is cross-ribbed (Fig. 2, next page). For obliquely ribbed bars, the diameters range from 6 mm to 28 mm (Fig. 3, next page). This last type of reinforcement is also called "tentor bar" and it is the strongest reinforcement steel available.

The standard length of reinforcement bars is 9 m.

- ADVANTAGES / DISADVANTAGES: Although the strength of circular bars is sufficient for all Rural Building purposes, it is advisable to purchase ribbed bars if they are available in the market. Ribbed bars are preferred because their rough surface texture provides a better grip to the concrete. This, along with their greater strength, allows the Rural Builder to space the ribbed bars wider apart, thus saving materials and reducing the total weight of the member.

Fig. 1

Fig. 2

Fig. 3

Fig. 4

STEEL WIRE NETTING

Fig. 5

EXPANDED METAL FABRIC

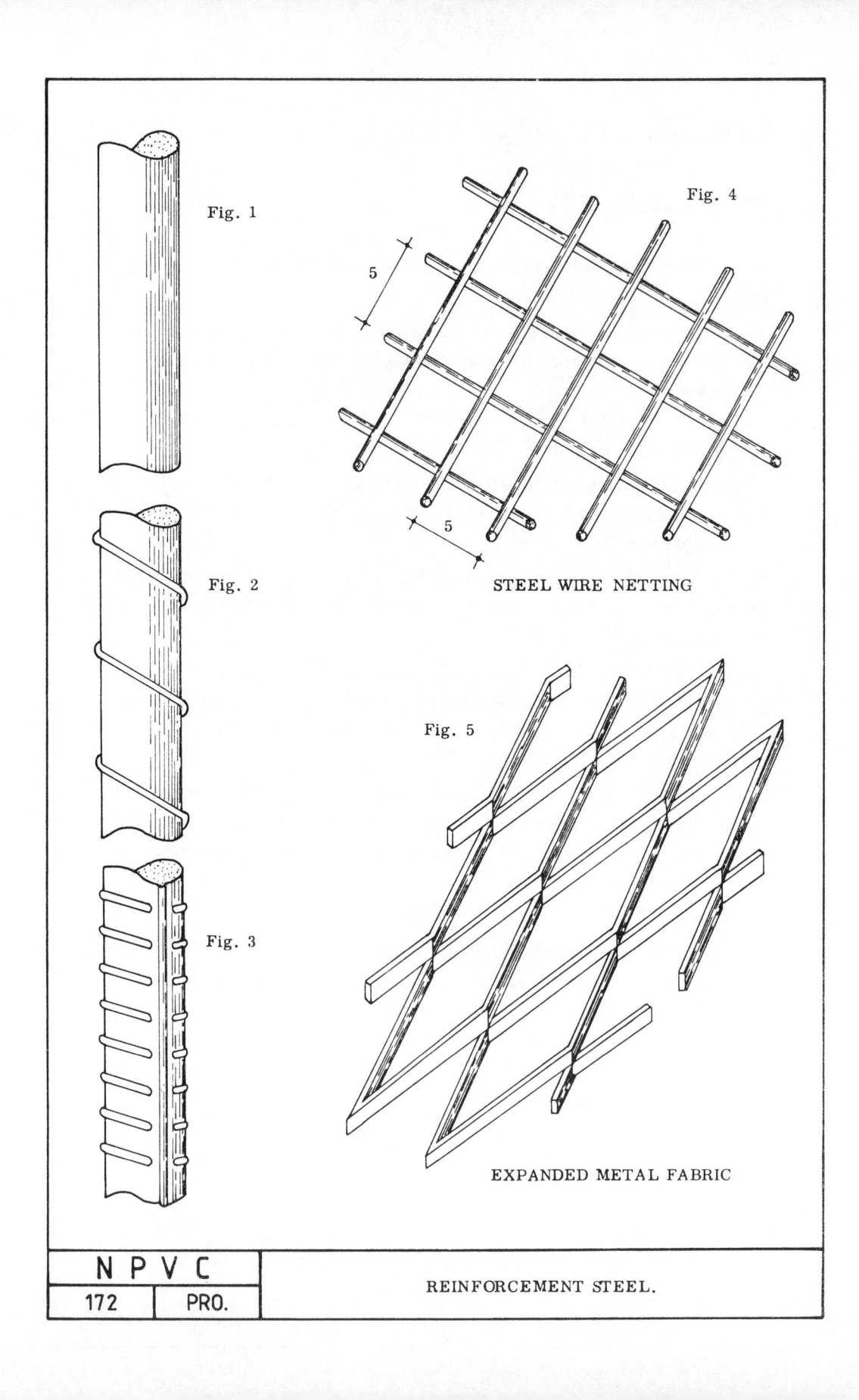

N P V C		
172	PRO.	REINFORCEMENT STEEL.

REINFORCEMENT MATS

A variety of reinforcement mats are available. They are usually made out of two layers of reinforcement bars laid across each other and secured together by welding.

The mats are either square or oblong in shape. They reduce the work needed to reinforce large members of the structure such as floors, walls, slabs, etc.

Regular reinforcement mats are hardly necessary in Rural Building, but two special kinds are frequently used for burglar proofing and to reinforce thin concrete slabs like manhole covers, draining boards in kitchens, and coping slabs.

These two are "expanded metal fabric" and "steel wire netting".

- STEEL WIRE NETTING: The most common steel wire netting (Fig. 4) has square meshes measuring 5 by 5 cm and is manufactured in the same way as reinforcement mats. The same kind of wire mesh can also have oblong meshes.

- EXPANDED METAL FABRIC: This is made by slitting metal sheets and then stretching them to form a diamond-shaped mesh (Fig. 5). Always wear leather gloves when working with expanded metal fabric, as the edges are very sharp.

Reinforcement mats are sold in sheets approximately 2,15 m wide and 5 m long.

Expanded metal fabric and steel wire netting can be purchased in sheets of about 1,5 m wide and 2,5 m long.

BINDING WIRE

This is a soft steel wire about 1 mm in diameter, used for binding reinforcement bars at the points where they cross each other. It is bought in rolls and may also be called lashing wire, annealed wire or tying wire.

NOTES:

LANDCRETE BLOCKS

One of the smallest but most important members of the structure is the block. Almost all walls in Rural Building are erected with blocks, preferable landcrete blocks.

- LANDCRETE: This word comes from the words laterite, land, and concrete.

    The land on which we live provides us with the laterite; the first syllable is a combination of the first two letters of laterite and the last two letters of land.

    Concrete as well as landcrete contains cement. In order to show this, the last syllable of the word concrete is used, making the word LANDCRETE.

Landcrete is a low cost, long-lasting and attractive building product. This chapter is about making landcrete blocks using a hand operated block press.

- LATERITE: This type of soil is found throughout the tropics. Its colour can vary from white-grey to a dark red, depending on the iron content. Laterite consists mainly of fine and coarse sand mixed with clay.

    Laterite has been used to make houses for a long time, but such walls break down easily and get washed away by rain. Pressing the soil into blocks makes it easier to build the walls, and they are stronger and more resistant to rain. By adding some cement to the laterite it is stabilized and makes even better blocks. The basic material, laterite, costs nothing and is usually found on the building site. It is easy to find good soil for building or to mix it with sand or clay to make it good.

The basic steps of the operation to make landcrete blocks are fairly simple: first good soil is found and tested; then it is prepared for the block press, with the addition of cement or lime if available. The soil or soil-cement is put into the press and compacted, raised out and removed for curing.

- NOTE: If the blocks do not contain cement they are not called landcrete; they are simply called "laterite blocks".

REQUIRED MATERIALS

a - Laterite soil: composed of sands, silt and clay
b - Water: to wet the soil; it should be clean
c - Cement: to stabilize the soil.

- NOTE: If you have no cement you can use lime (twice as much as the amount of cement recommended) or else just make plain laterite blocks.

REQUIRED EQUIPMENT

a - Block press with proper mounting rails and a wooden handle (2,5 m long)
b - Box for shrinkage test
c - Headpan, box or bucket for batching
d - Pick-axes and shovels for digging, mixing and filling.

TESTING AND CHOOSING THE SOIL

Most soil is suitable for making blocks, but it must be tested first to find out how much sand, silt and clay it contains.

Dig a small pit for testing. First remove and set aside the top soil where plants or grass may be growing (25 to 50 cm deep). This soil should not be used for blocks. Dig out the soil under the top soil. The deeper soil may be sandier, which is usually better for making blocks.

Now make three tests: a - Drop test   b- Jar test   c - Box test.

- **DROP TEST**: Take a handful of soil which is wet enough to form a ball, and squeeze it in your hand, but not so tightly that the water is squeezed out.

Drop the ball from about one meter high onto hard ground. If it breaks up into only a few pieces, the block-making quality is good. If it breaks completely up, there is either not enough water in it or not enough clay, and the quality is bad.

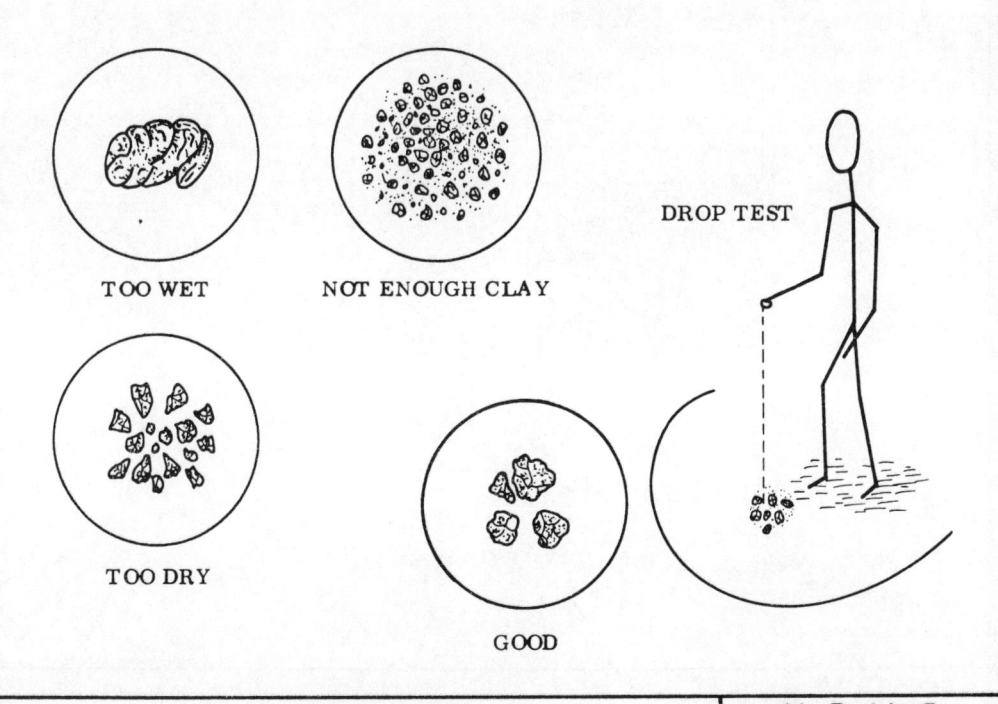

TOO WET

NOT ENOUGH CLAY

DROP TEST

TOO DRY

GOOD

LANDCRETE BLOCKS.

N P V C
PRO.   175

Fig. 1

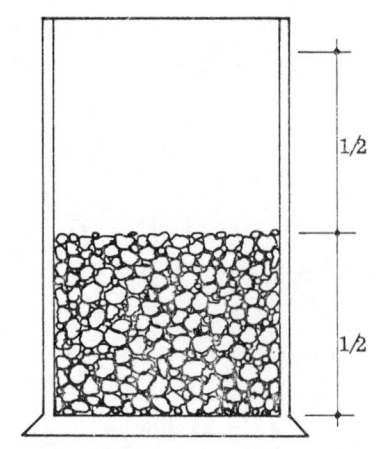

FILL GLASS JAR HALF FULL WITH LATERITE SOIL

1/2
1/2

Fig. 2

**FILL WITH WATER AND ADD A LITTLE SALT
SHAKE IT THOROUGHLY !**

Fig. 3

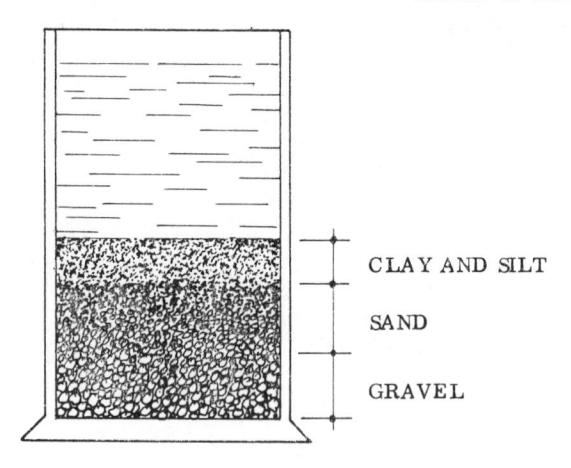

CLAY AND SILT
SAND
GRAVEL

USE SOIL THAT CONTAINS:
5 - 30% CLAY AND SILT
AND AT LEAST 30% SAND

N P V C		LANDCRETE BLOCKS.
176	PRO.	

- JAR TEST: This test separates the sand from the clay and silt, so that we can measure the quantity of each.

First dig out some soil (not top soil). Fill a glass jar half-full with the soil (Fig. 1).

Fill the jar up with water and add two teaspoons full of salt, to make the particles settle faster (Fig. 2).

Cover and shake or stir the jar for two minutes to mix the water thoroughly with the soil.

Set the jar on a level surface and leave the soil to settle for several hours. Sand and gravel will settle to the bottom, leaving the silt and clay on top (Fig. 3).

Measure the height of each layer to find the total amount of sand and the amount of clay and silt, compared to the total amount of soil.

The soil you use should contain at least 30% sand (about 1/3), and between 5% and 30% clay and silt. If there is not enough clay add more or else find some better soil. You can also add sand if necessary.

NOTES:

LANDCRETE BLOCKS.

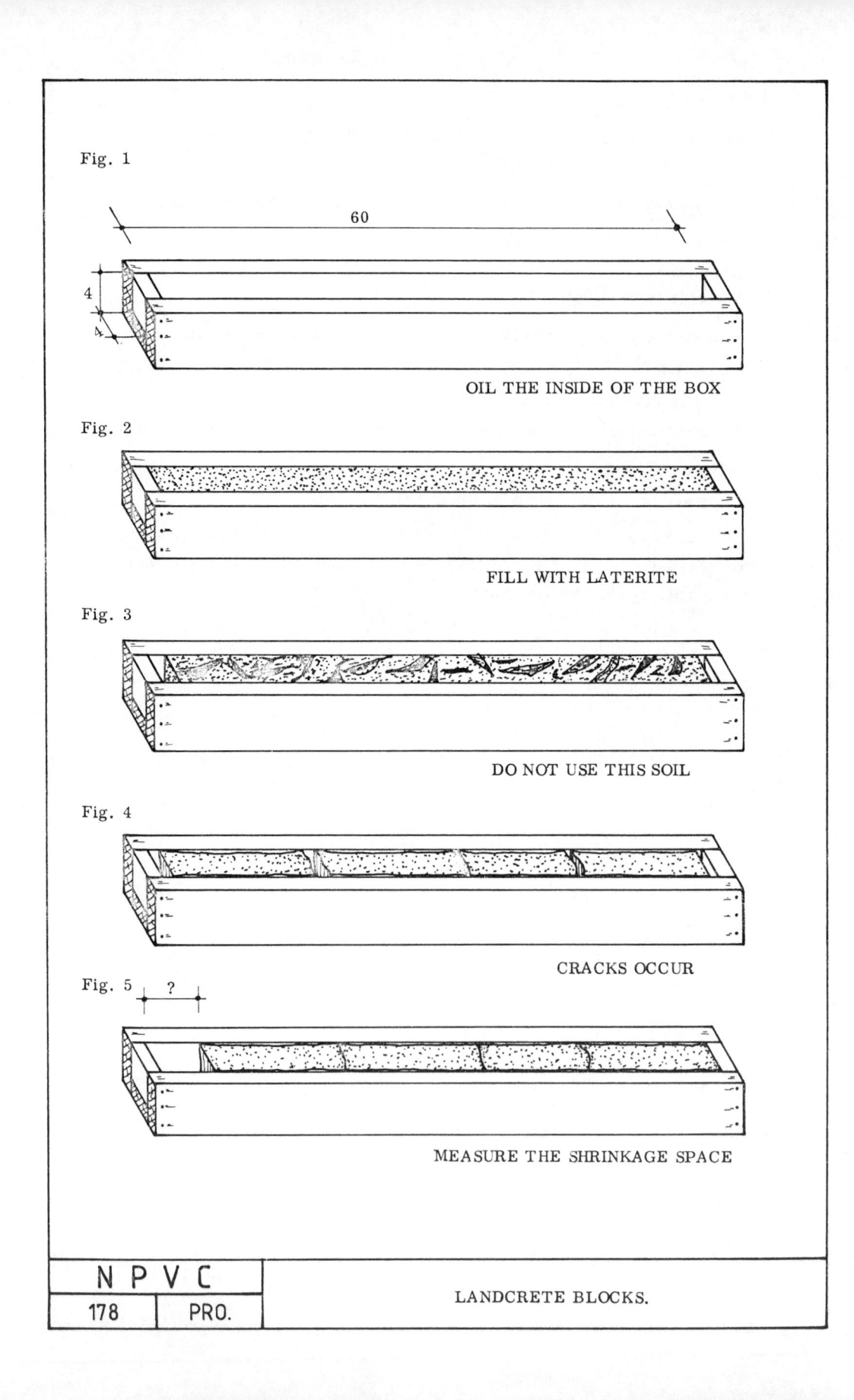

Fig. 1

60

4
4

OIL THE INSIDE OF THE BOX

Fig. 2

FILL WITH LATERITE

Fig. 3

DO NOT USE THIS SOIL

Fig. 4

CRACKS OCCUR

Fig. 5 ?

MEASURE THE SHRINKAGE SPACE

N P V C	
178	PRO.

LANDCRETE BLOCKS.

- **BOX TEST:** This test shows the quality of the soil and allows you to determine the amount of cement you should use with it.

Use an open wooden box with inside measurements of 60 cm by 4 cm by 4 cm (Fig. 1). Oil or grease the inside of the box.

Fill the box with very wet soil. Compact it well, especially in the corners, and level off the top with a stick or the edge of your trowel (Fig. 2).

Put the box in the sun for three days to dry, or in the shade for seven days. It should be protected from rain.

The soil will shrink as it dries. Do not use soil for blocks if it has many cracks in it (Fig. 3) or if it has arched up out of the box. Don't use soil if it has shrunk more than 5 cm. Either find some better soil, or improve the soil by adding sand, since it is the clay which causes shrinkage.

Measure the shrinkage by tapping one end of the box on the ground so that all the soil slides down to one end (Fig. 4). The cracks will close and you can measure the shrinkage space at the top end (Fig. 5).

The amount of shrinkage tells you how much cement you should use. The more shrinkage, the more cement is needed. Use the table below as a guide for the amount of cement to be added to the laterite.

- **TEST TABLE:**

Shrinkage	Cement to soil
0 - 10 mm	1 : 35
10 - 20 mm	1 : 30
20 - 30 mm	1 : 25
30 - 40 mm	1 : 20
40 - 50 mm	1 : 15

NOTES:

Fig. 1

TOP SOIL
PIT
SOIL — 50 cm
— 100 cm
SANDY SOIL
GRAVEL — 150 cm
— 200 cm
ROCKS
— 250 cm

N P V C	
180	PRO.

LANDCRETE BLOCKS.

MAKING BLOCKS

- **PREPARATION OF THE MIXTURE:** After you have found good soil and the correct amount of cement to use with the help of the box test, the soil mixture must be prepared for the block press.

If you have no cement and must make laterite blocks, you follow the same sequence as described below.

a - Remove the top layer of soil (Fig. 1).

b - Dig out the soil you want to use and pile it (Fig. 1).

c - Measure the required proportions of laterite and cement.

d - Make a dry mix of the batch.

e - Add water and make a wet mix.

f - Check the moisture content using the drop test.

If you are making laterite blocks, steps c and d are of course left out.

Before you start batching, the laterite must be broken up so that no lumps remain. This is usually done by beating the soil with the back of a shovel or with a piece of wood. Large stones are removed.

- **MIXING:** Use flat, hard ground for mixing. If no such place is available, prepare a mixing platform before you start working.

Spread the laterite out until it is about 10 cm thick. Spread the cement evenly over all the soil. Mix the cement and soil with a shovel until the mixture is of an even colour throughout (about 3 times - see Rural Building Materials, page 163).

Spread the heap again, sprinkle a little water over it and mix. At this point, test the mixture for moisture with the drop test. If it is too dry, spread it out again and add more water.

The soil-cement mixture is now ready for the block press. There should be enough mix for about 7 or 8 blocks in one batch at a time.

- **NOTE:** Never prepare more than you can use up within 30 minutes (about three batches). It is better to mix small amounts more often.

NOTES:

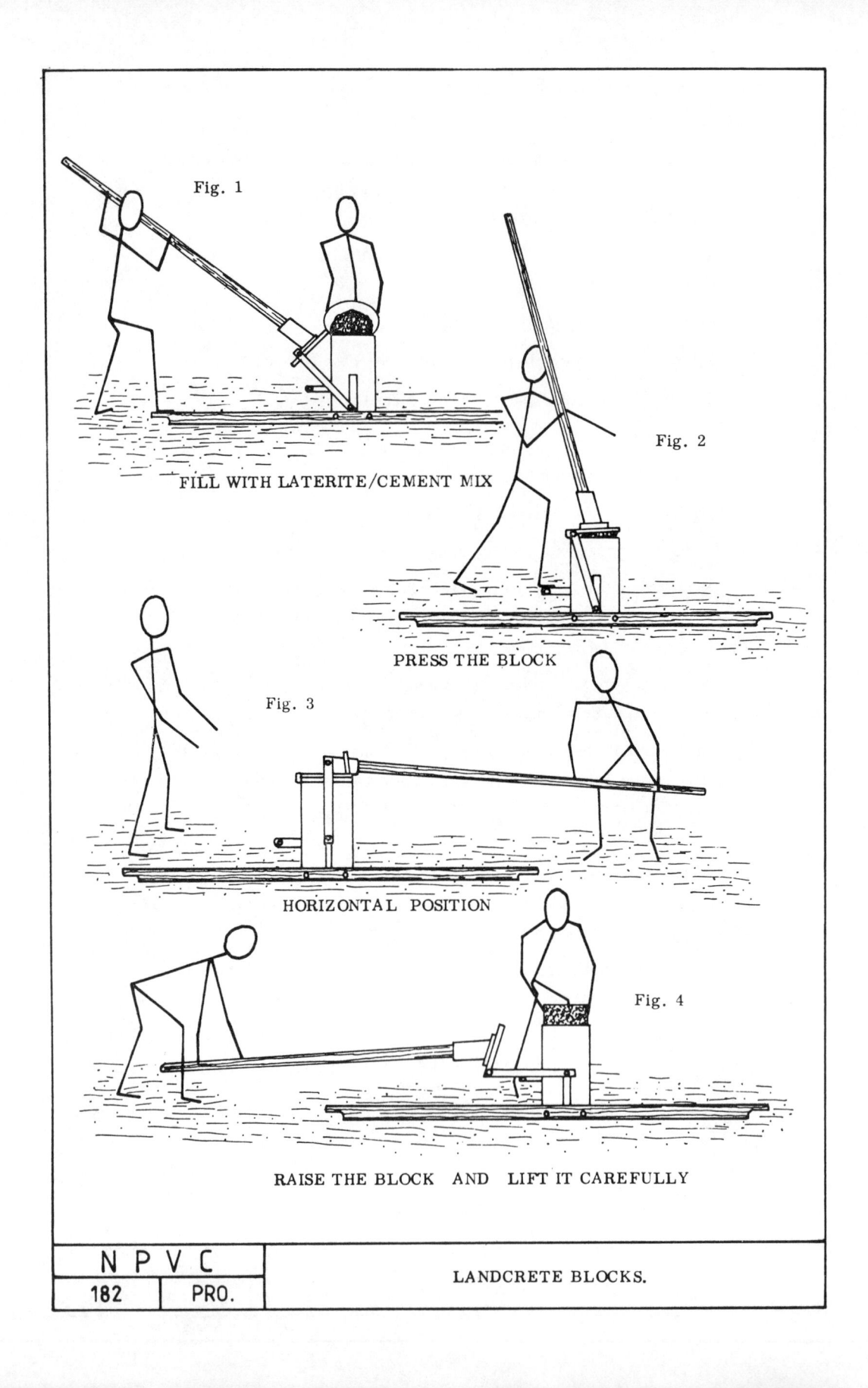

Fig. 1

FILL WITH LATERITE/CEMENT MIX

Fig. 2

PRESS THE BLOCK

Fig. 3

HORIZONTAL POSITION

Fig. 4

RAISE THE BLOCK AND LIFT IT CAREFULLY

N P V C		
182	PRO.	LANDCRETE BLOCKS.

- PRESSING THE BLOCK:

    a - Place the block press in its rails on flat, solid ground near the mixing platform.

    b - Open the mould box by swinging the handle down to the lower roller.

    c - Half-fill the mould box with the laterite cement mix (Fig. 1).

    d - Press the mix firmly into the corners with a piece of wood.

    e - Fill the mould to the top and compact the corners again.

    f - Add a little more so that the mould is filled flush to its top edge.

    g - Swing the handle quickly over to the other side and press the block until the handle has reached a horizontal position (Figs. 2 & 3).

    If the mix is too dry, the handle will not go all the way down to the horizontal position. In this case do not force it, as the handle may break.

    Instead, eject the unfinished block so that you can refill the mould box after adding a little water to the mix.

    On no account should more than one man at a time work the handle!

    h - Raise the block out of the mould box by swinging the handle back against the lower rollers (Fig. 4).

    i - Lift the block carefully off the machine and place it for drying.

The freshly made block is not strong yet. If it breaks or cracks very easily, the mixture is not correct. Try a different mixture.

Hold the block in such a way that your fingers are not caught under it when you put it down, so as not to crumble the edges.

NOTES:

LANDCRETE BLOCKS.

Fig. 1
DRYING FOR ONE DAY

Fig. 2
STACKING FOR FURTHER CURING

BLOCKS WITHOUT CEMENT – SUN DRYING

Fig. 3

N P V C	
184	PRO.

LANDCRETE BLOCKS.

- CURING: This is the term originally used to describe the chemical change in glues when they set, meaning when they become strong and hard.

As far as cement products are concerned, curing simply means the after-treatment of any of those products.

If the blocks contain cement, we talk about "curing". If the blocks don't contain cement, we talk about "sun-drying".

The blocks with cement must now be cured for about two weeks while the cement sets. It is important to follow the directions for curing. If you do not, the blocks may be weak and full of cracks, and therefore unusable for building.

a - Remove the block from the block press, holding it carefully.

b - Place it on leaves, grass or boards on flat ground. The block should not touch the ground (Fig. 1).

c - The blocks should be under shelter or covered with something so that they are out of the rain and sun for at least the first day.

d - Let the blocks dry like this for one day.

e - After one day the blocks are a little stronger so that they can be stacked for further curing.

Stack the blocks on boards or on very flat, hard ground up to five blocks high. Place them so that they touch each other. Make the stacks under a cover if possible, to keep them out of the sun (Fig. 2).

f - The blocks must be kept moist by sprinkling water on them twice a day. Put grass or leaves on top to help keep them moist (Fig. 2).

g - After two weeks of watering the cement has set properly and the blocks can dry completely. They are now ready for use.

Blocks without cement simply need to be dried in the sun. Let the blocks dry in the sun for two weeks; then they are ready for building (Fig. 3).

NOTES:

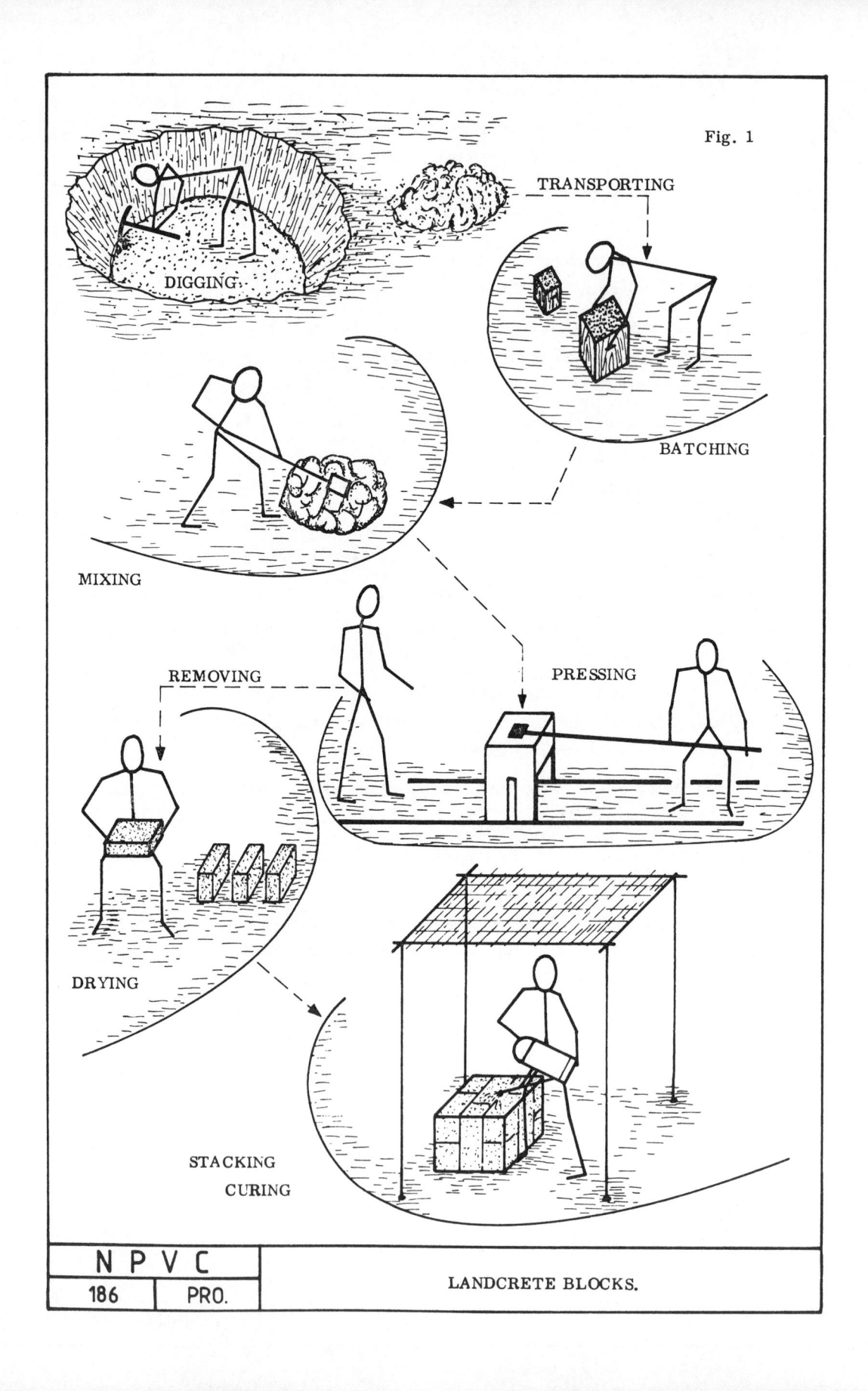

Fig. 1

DIGGING

TRANSPORTING

BATCHING

MIXING

REMOVING

PRESSING

DRYING

STACKING
CURING

N P V C		LANDCRETE BLOCKS.
186	PRO.	

PLANNING THE WORK

- **PLAN OF OPERATIONS:** Good planning can make the work of block-making go faster and easier.

The places where the different steps are carried out should be as close to each other as possible, so that there is a continuous step by step flow of laterite from the soil pit to the finished wall. The soil and blocks should be transported as little as possible.

If there is more than one building to erect, some operations can be moved.

There should be a smooth flow (Fig. 1) of:

a - digging
b - transporting
c - batching
d - mixing
e - testing
f - filling
g - pressing
h - raising
i - removing/drying
j - stacking/curing
k - walling up

You can make a layout like the one in the picture (Fig. 1) or you can lay out the operations in a straight line; or anything in between, whatever suits you and the situation best.

Several factors can be important in deciding what sort of digging pits you will have. If the good soil goes deep, all the soil can come from one pit. However such a large pit might be ugly and undesirable; several small pits could be a better solution in some cases.

The possible future uses of the pits should also be considered. They could form a part of a drainage system, a water storage tank, a sewage pit, a soak-away, and so on. This of course, provided that the planning is done beforehand.

NOTES:

Fig. 1

NUMBER OF MEN							KIND OF ACTIVITY
ONE MAN ONLY				2	3	4	DIGGING
	1	2		2	2	3	TRANSPORTING
			1	2	2	1	BATCHING AND TESTING
						2	MIXING
				1	1	1	FILLING AND OILING
			1			1	PRESSING
					1	1	RAISING OUT
				1	1	2	REMOVING AND PLACING TO DRY
1	2	3	6	8	10	15	TOTAL MANPOWER

Stacking, curing and transporting the blocks to the actual place of building can be done every morning by the soil diggers.

They can do this only if they have dug some soil in advance during the previous day, so that the rest of the workers have the materials to continue their operations of batching, mixing etc.

NPVC 188 PRO.   LANDCRETE BLOCKS.

- LABOUR: From one to fifteen men can work on block-making. If there are plenty of workers, they should be organized to keep the block press going constantly, so it is used to maximum efficiency.

To do this, there must be a steady supply of soil-cement mix ready to put into the machine. Make the mixing platform big enough so that there is room for one pile of already mixed landcrete and one pile which is being mixed.

You can use the rough table at left as a planning guide to divide the labour, but experience will be the best guide (Fig. 1).

In any case the work should be divided so that everyone is busy all the time. If the block press filling worker has to stop and wait for prepared landcrete, he or another man should be switched to doing soil preparation.

Workers should relieve one another in their jobs every few hours to prevent boredom with the work. After a few days of such rotations, the workers will each become skilled and efficient at three or four of the different steps of block-making.

It is important to share the work fairly to keep up the morale and enthusiasm among the workers.

- NUMBER OF BLOCKS: You should know from the start approximately the number of blocks that will be needed for the building. This is necessary to be able to schedule the block-making, curing and building.

To find the approximate number of blocks to be made, you must know the size and plan of the building. Take measurements of the lengths of the walls and add these up to get a total. Multiply this wall-length by the total height from the plinth course to the top of the wall; this gives you the total wall area in square meters. This number, multiplied by 13,5 (the approximate number of blocks per square meter) gives you the total number of blocks to be made.

- REMEMBER: Wall area in square meters x 13,5 = Number of blocks. (Also, see the Tables of Figures, pages 237 and 238).

NOTES:

LANDCRETE BLOCKS.

Fig. 1

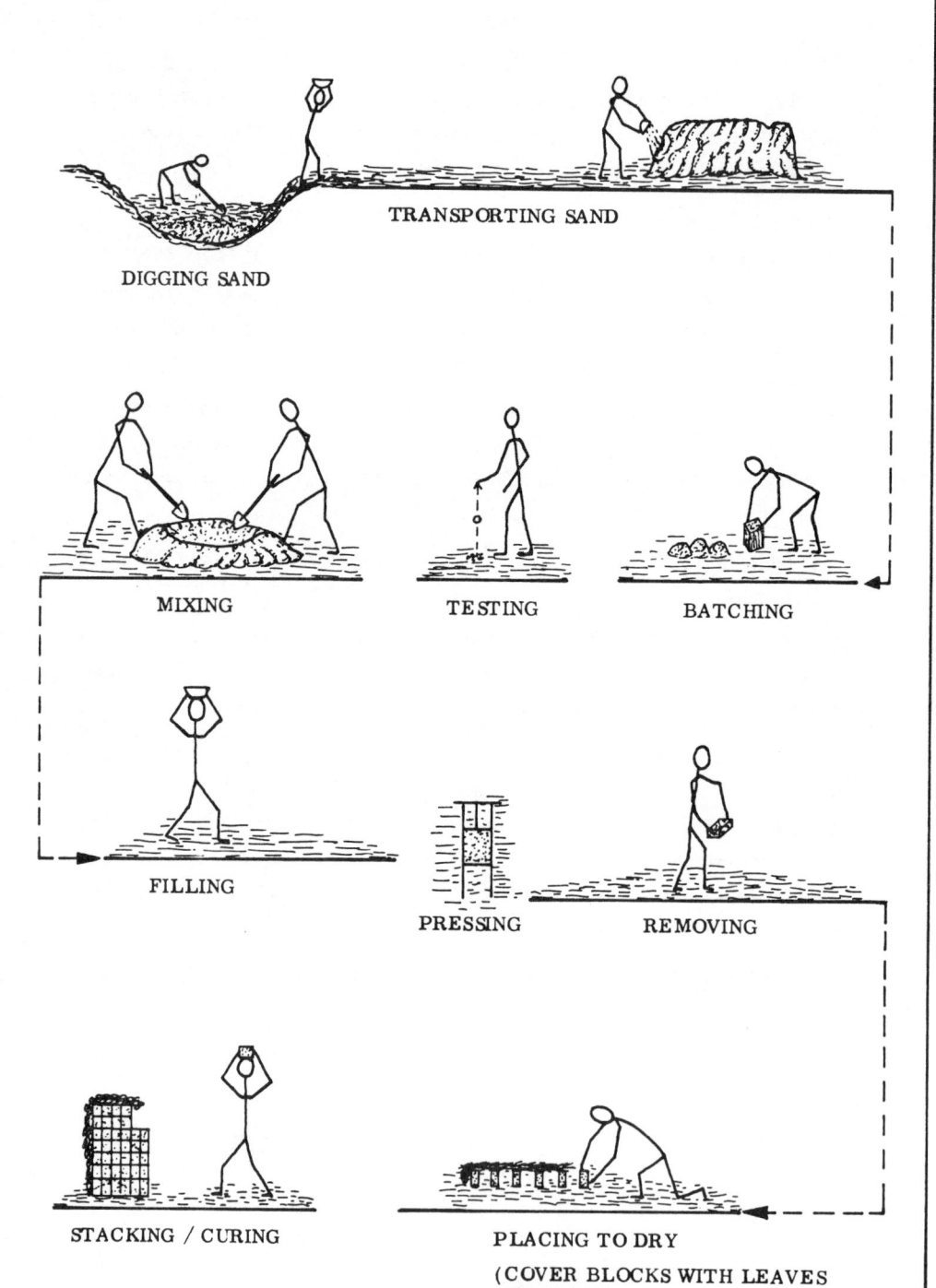

DIGGING SAND

TRANSPORTING SAND

MIXING

TESTING

BATCHING

FILLING

PRESSING

REMOVING

STACKING / CURING

PLACING TO DRY
(COVER BLOCKS WITH LEAVES OR PAPER)

N P V C		
190	PRO.	SANDCRETE BLOCKS.

SANDCRETE BLOCKS

The term sandcrete comes from "concrete" by replacing the first syllable "con" with the word "sand". This is done to make it clear that this building product contains only sand as an aggregate, and no stones. It can also be called "fine-grained concrete" but the new term sandcrete is preferred as it corresponds to landcrete.

A pallet is put into the mould box of the machine (see page 29) and the box is filled with a mixture of cement and sand; then the lid of the machine is used to compact the material to the required size (for proportions of the mix, see the Tables of Figures, page 234).

Unlike landcrete blocks, sandcrete blocks have to be made upon a pallet, as they are too soft to be carried when freshly made. Differently shaped blocks can be made with this machine by changing the height of the pallet or by using inserts.

MAKING THE BLOCKS

Making blocks with this machine is similar to making landcrete blocks.

a. Put one or more pallets into the mould box, according to how thick you want the blocks.

b. Half-fill the mould box with the ready mixed sandcrete.

c. Compact the corners with a piece of wood.

d. Fill the mould box completely and again compact the corners. Add a little more if necessary to fill the box flush to the top edge.

e. Compact the sandcrete by repeatedly banging the heavy lid on it, until the lid fits exactly in its lowest position. Sometimes the lid does not close properly because the mould box is too full. In this case, scrape off a small amount of sandcrete with your trowel and repeat the compaction. If you fail to do this the block will be wedge-shaped and difficult to set in the wall.

f. Open the lid wide and pull the handle to push the block out.

g. Remove both pallet and block at the same time and set them in place for hardening and curing.

- NOTE: Before use, the pallets must be soaked in water thoroughly, to prevent them from bending during the drying process. If this is not done the pallets will probably bend and crack the blocks.

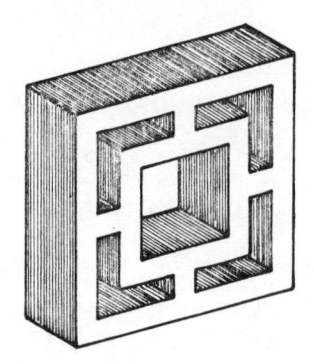

Fig. 1

Fig. 2

Fig. 3

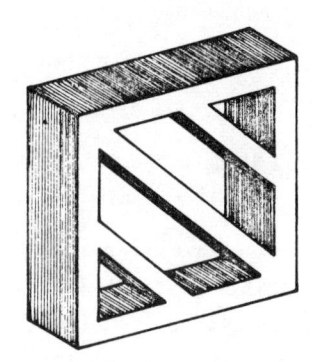

Fig. 4

Fig. 5

NPVC | 192 | PRO.

DECORATIVE BLOCKS.

PLANNING THE WORK

As in the plan of operations for making landcrete blocks (page 187), for sandcrete blocks also the speed and ease of the work depend on how well it is planned.

Fig. 1 on page 190 shows how the block-making can be planned.

DECORATIVE BLOCKS

Originally, a decorative block was understood to be a solid block with decorative textured faces. What we now commonly call decorative block is in fact part of a decorative openwork screen built into an opening. The correct term is "decorative grille" (also spelled "grill").

This kind of block is made in a special iron mould. It can serve several purposes:

- To give an attractive appearance
- To provide light without installing burglar-proofing or any kind of louvres, shutters, etc.
- To provide permanent ventilation without using ventilation blocks
- Or a combination of two or three of the above requirements.

The illustrations of blocks on the opposite page show that almost any design is possible, given a fertile imagination. Remember however that the strength of the blocks depends also on their shape (Figs. 1 to 5).

- NOTE: To make it easier to empty the mould, short pins can be welded onto each corner at the top of the mould. This allows you to tap the pins gently on a hard, level surface; thus loosening the block from the mould (Fig. 6).

Drilling small holes into the bottom of the mould can also make it easier to remove the block. The holes allow air into the mould as the block comes out.

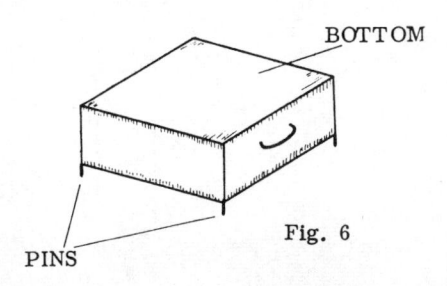

Fig. 6

Fig. 1

Fig. 2

Fig. 3

Fig. 4

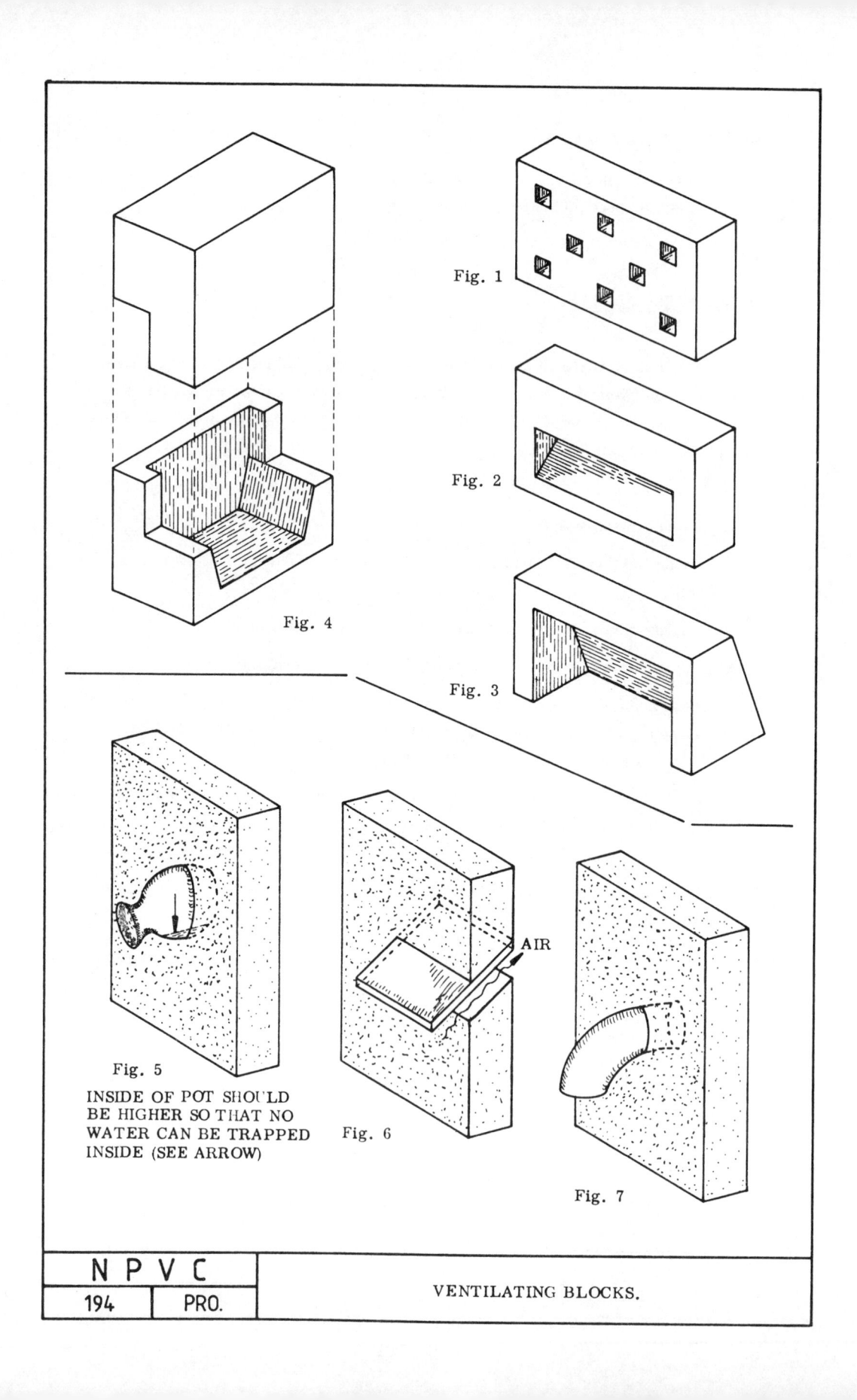

Fig. 5
INSIDE OF POT SHOULD BE HIGHER SO THAT NO WATER CAN BE TRAPPED INSIDE (SEE ARROW)

Fig. 6
AIR

Fig. 7

N P V C		
194	PRO.	VENTILATING BLOCKS.

VENTILATING BLOCKS

These are blocks which have an opening (or several openings) in them. They are used to ventilate rooms, stores, the spaces above ceilings etc.

There are various types of ventilating blocks. Some are designed to keep out rain, others include mosquito-proofing or a decorative front face (Figs. 1, 2 & 3).

In cases where a maximum amount of ventilation is desired, it is advisable to make a ventilating unit which is constructed out of two identical halves (Fig. 4). The inside of this unit can be painted in a bright colour, to increase the amount of light inside the room.

Simply shaped blocks for ventilation can be made in the sandcrete block machine by inserting wooden blocks according to the desired shape.

More complicated designs usually require a specially made wooden mould. The advantage of this kind of mould is that any size and shape of block can be made.

Pre-cast sandcrete or concrete are not the only choices of materials for ventilating units. Local potters' skills in baking earthenware can be used and one can design ventilating units from clay. Existing clay shapes can be used, such as tiles and pots (Figs. 5 & 6) or new shapes can be invented (Fig. 7).

When designing these ventilating units keep in mind the direction of the driving rain. Make sure that the inside of the unit is higher than the outside, and that there is no place for water to become trapped inside the unit to make a breeding place for mosquitos (Fig. 5).

Apart from the above considerations, there is no limit to the imagination of the Rural Builder in designing different shapes and kinds of ventilating units.

NOTES:

HOLES TO FIX WIRES

Fig. 1

ARRANGEMENT OF SHUTTERING BOARDS

SHUTTERING

Fig. 2

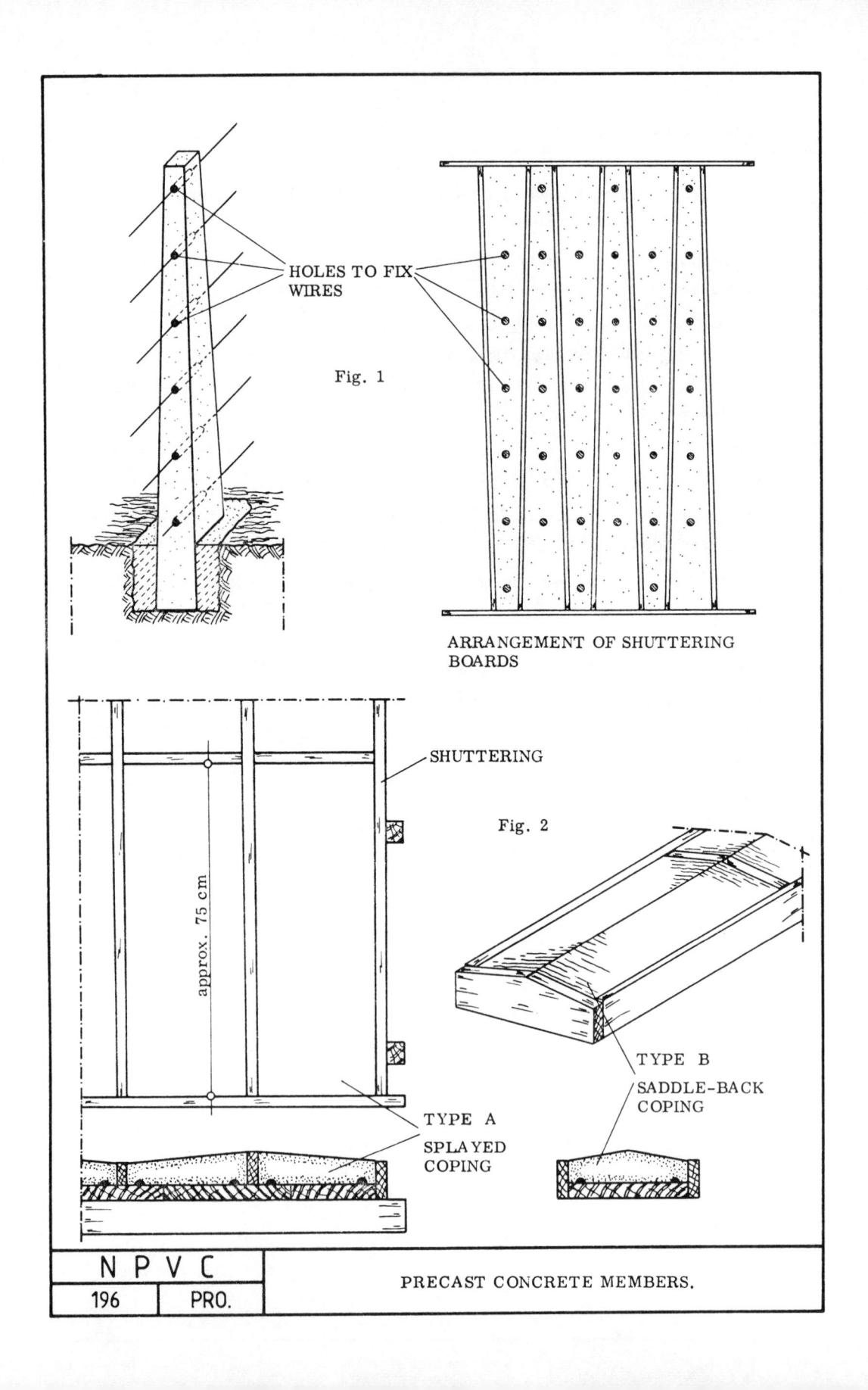

approx. 75 cm

TYPE A
SPLAYED COPING

TYPE B
SADDLE-BACK COPING

N P V C		PRECAST CONCRETE MEMBERS.
196	PRO.	

PRECAST CONCRETE MEMBERS

Many of the components of a building are prefabricated. This means that they are made in advance at any time and place, and can be used for any building. The most common examples are the various sorts of blocks, plywood, roofing sheets etc.

In the same way, many reinforced concrete members of the structure can be made well in advance so that they are ready to be used as the construction proceeds and they are needed. These are referred to as "precast" members, as opposed to the "cast-in-situ" members.

Generally speaking, prefabricated construction is divided into two classes with regard to buildings:

a - Prefabricated units produced in a factory and transported to the building site
b - Units produced by the contractor in a yard next to the building site.

Since Rural Building is understood to be conventional building and not factory production, the first method does not apply to us.

It is becoming more and more common however for the contractor to prefabricate members near the building site, although the possibilities of this method are limited to small-scale applications.

In Rural Building the most frequently made precast concrete members are:

- REINFORCED CONCRETE LINTELS: These cannot exceed 1,5 m in their length, as otherwise they will be too heavy and impossible to set in place without using a lifting device.

- REINFORCED CONCRETE POSTS: For fencing purposes (Fig. 1).

- REINFORCED CONCRETE PILLARS: These should not be too heavy to be set up under rural conditions.

- CONCRETE COPINGS OR CAPPINGS: These are of various shapes, with or without reinforcement (Fig. 2).

- REINFORCED CONCRETE SLABS: Of limited size and thickness, used to cover manholes or to serve as draining boards in kitchens, etc.

NOTES:

PLANNING THE WORK

All the plans and if possible, detailed drawings must be available before you start to make precast concrete members. Careful planning, supervision and performance are all necessary in order to obtain the desired product.

- SAVING TIME: As soon as the required materials are available, a group of the workers can start production of precast concrete members. The earlier they begin the better, because the concrete needs to be cured for some time before it is used in the building. By the time the precast member, for example a lintel, is needed the curing process will be complete and the lintel can simply be set into place.

  In contrast to the cast-in-situ method, this method allows construction to proceed without delays caused by waiting for concrete to harden; and without being hindered by shuttering and strutting. In this way the total construction time is shortened.

  This time-saving can become very important, for example if the wet season is approaching and the building has to be done before that time.

- SAVING MATERIALS: Since precast concrete members can simply be made on levelled ground, there is no need for poles and braces or other strutting to hold up the **shuttering**. Curing is also made much easier.

  In addition, the same formwork can be used repeatedly if several members of the same size and shape have to be made.

  If there is no proper storage space for the cement on a job, it is better to precast as many members as possible; thereby using up the cement before it gets spoiled.

- NOTE: The more time, materials and money you save, the happier your client will be. Not only that, but your reputation as a Rural Builder will grow in the eyes of the people around you and be appreciated in the society.

NOTES:

GLUE

Glue is a liquid which is used to stick materials together, particularly on wooden surfaces.

In Ghana's Upper Region, the Rural Builder cannot use glue very often, because of the problems caused by the difference in humidity between the dry and rainy seasons. Glue is sometimes applied to make joints stronger, but it should not be used alone. There must always be some additional fastening; either screws or nails. To rely on joints fastened simply with glue is asking for trouble.

Three points should be kept in mind when you use glue:

a - The parts to be glued have to be in close contact.
b - There must be a large area of contact for glueing.
c - End-grain does not glue well, so a joint which consists mostly of end-grain will be weak.

There are many different glues but in Rural Building we use only two kinds; synthetic glue and contact glue.

- SYNTHETIC GLUE: The most common synthetic glue is PVA (polyvinyl acetate) glue. It is a milk-white glue.

 The parts to be glued should be clean and well-fitting. The parts are squeezed together immediately after the glue has been applied, and clamped together until the glue has set hard. The joint will have considerable strength after the glue has set for one hour. PVA glue is not waterproof. If the joint must be waterproof, special kinds of glue have to be used.

- CONTACT GLUE: This glue is used primarily to glue laminated plastics such as Formica or similar materials to sheet materials.

 The glue is spread evenly over both surfaces. A toothed spatula can be used for this purpose. After fifteen minutes the surfaces are pressed together. It is important to take care that the surfaces are exactly in the correct position when they are put together. Once the surfaces stick they cannot be separated again for adjustment.

Take care that no air is trapped between the two surfaces when they are glued, because afterwards it will be impossible to get the air bubbles out.

PROTECTIVE FINISHES

In Rural Building we have the following finishes which are applied on wood or masonry work:

- Oil paint
- Synthetic paint
- White wash
- Cement paint
- PVA Emulsion paint
- Varnish

OIL PAINT

The traditional type of oil paint has a vegetable oil base (linseed oil) and a pigment which gives it colour.

Oil paint is usually applied in three coats. Each coat is of a different composition and they cannot be mixed.

- Priming coat
- Undercoat
- Finishing coat

Follow the directions on the tin when using these paints. Oil paints can be diluted only with thinners, such as turpentine.

SYNTHETIC PAINT

Synthetic paints have a chemical base and a pigment. These paints dry more quickly than oil paints and they are more weather resistant.

Read the directions on the tin before using the paint.

WHITE WASH

White wash is often used for interior work. It is composed of lime and water and it is not water resistant. Its lack of water resistance and its poor wearing qualities make it inferior to emulsion paint as a finish for outside surfaces.

White wash is often used as a priming coat for emulsion paints. It fills in the pores in cement or plaster and makes the surface smoother so that less emulsion paint is needed.

After mixing the lime with water it is advisable to leave the white wash for a day to stand, to be sure that no more chemical reactions are taking place. Follow the manufacturer's directions on the label when mixing the white wash.

CEMENT PAINT

Cement paints are often used externally. They contain white or coloured Portland cement and are sold in powder form. This paint should be made workable with water only; when dry it forms a waterproof seal on the concrete or masonry.

PVA EMULSION PAINT

Polyvinyl acetate emulsion paint has a latex (rubber) base and a pigment for the colouring. This paint is used mostly for internal and external masonry work.

The latex paint can be diluted with water if necessary to improve its workability. The paint should be applied in thin layers and the directions on the label should be followed. Read them before you start work.

VARNISH

Varnishes are used to protect wood. There are two kinds of varnish: oil or spirit. Oil varnish can be used for external work. It is diluted if necessary with turpentine.

Spirit varnish is only used for internal work such as for furniture. This varnish is not very strong or water resistant. Spirit varnish can be diluted with commercial alcohol.

PAINTS AND VARNISHES: PURCHASING

Paints and varnishes are sold in containers of one or more litres. On each container there should be a description of how to apply the paint or varnish. There should also be a date stamped on the container, to indicate how long the paint or varnish will last (when it will be too old to use any longer).

NOTES:

Fig. 1 PLYWOOD

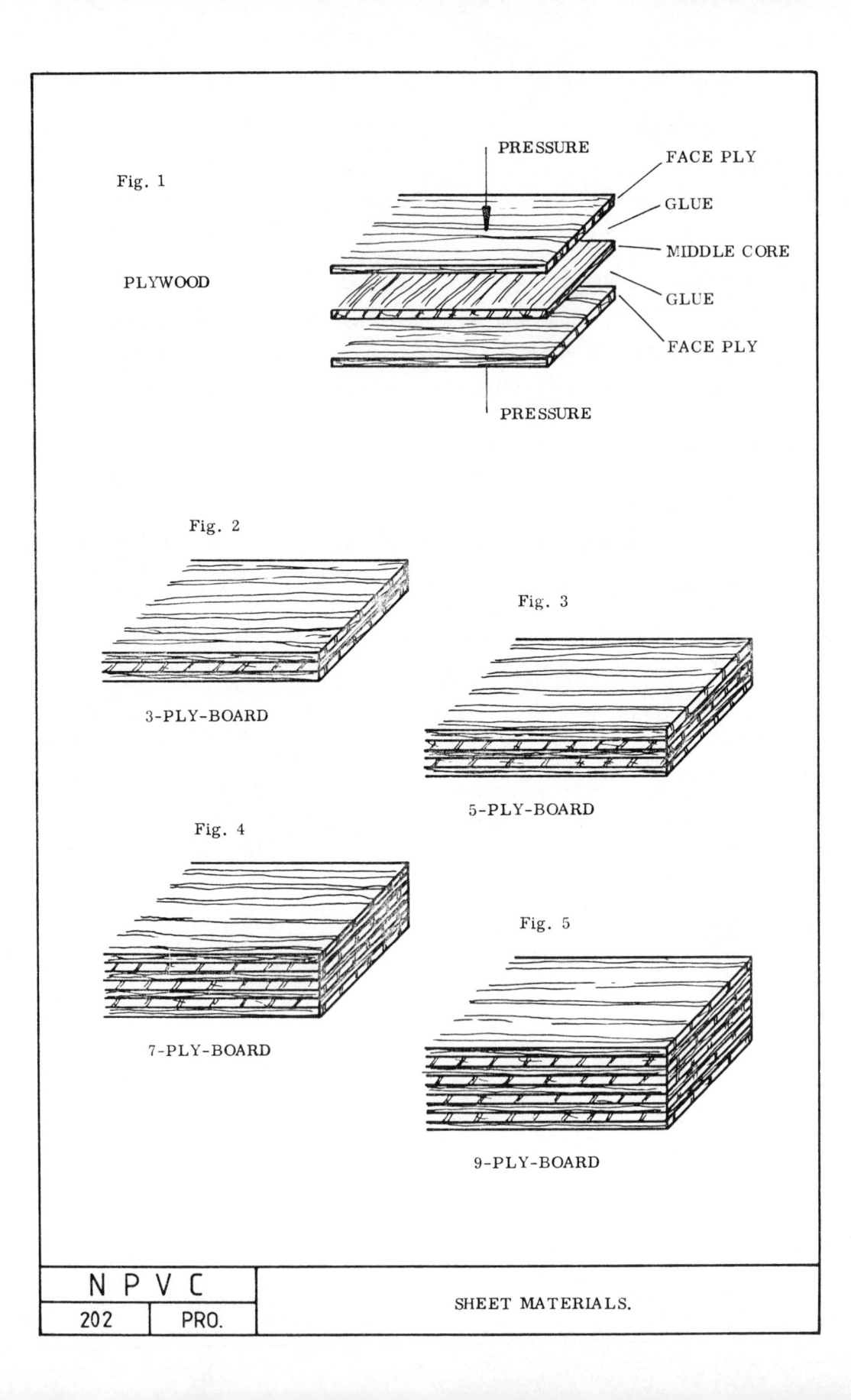

PRESSURE
FACE PLY
GLUE
MIDDLE CORE
GLUE
FACE PLY
PRESSURE

Fig. 2 3-PLY-BOARD

Fig. 3 5-PLY-BOARD

Fig. 4 7-PLY-BOARD

Fig. 5 9-PLY-BOARD

N P V C 202 PRO. SHEET MATERIALS.

SHEET MATERIALS

Even as new and wonderful materials are becoming available to the Rural Builder, timber is still in very great demand. Wood is easy to work with, adaptable and durable when cared for properly.

Sheet materials manufactured from sawmill wastes are used more and more, both because they save money and because they do not have some of the problems with shrinking or splitting that affect natural materials.

The sheet materials we deal with in Rural Building are: plywood, blockboard, hard board, chip board and decorative laminated plastics.

PLYWOOD

Plywood manufacture is the oldest means of improving the properties of timber. Large sheets can be made, free from defects and unaffected by shrinkage and splitting.

The plywood is made by glueing together several thin layers, called plies or veneers, so that the grains of each run crosswise to its neighbours. There is always an odd number of plies so that the grains of the two outer layers run in the same direction. This is so that the plywood remains flat (Fig. 1).

Plywood is so useful because of its special properties:

- It is stable and will not expand or shrink like solid timber; however it will absorb moisture and may tend to curl as the surface layers expand a bit.

- It is very strong because of the crossed grain structure. Even the thinnest plywood cannot be split.

The number of plies can be from 3 to 9, making sheets which are 3 to 25 mm thick (Figs. 2 to 5).

NOTES:

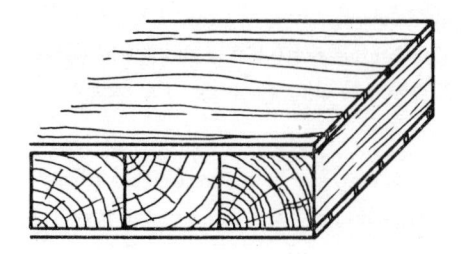

Fig. 1

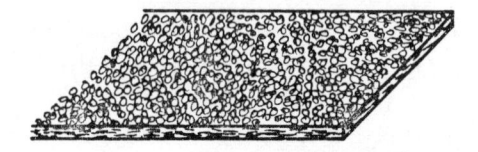

Fig. 2

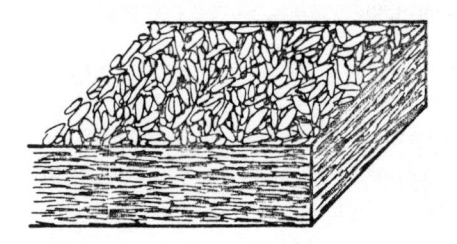

Fig. 3

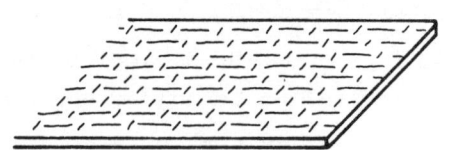

Fig. 4

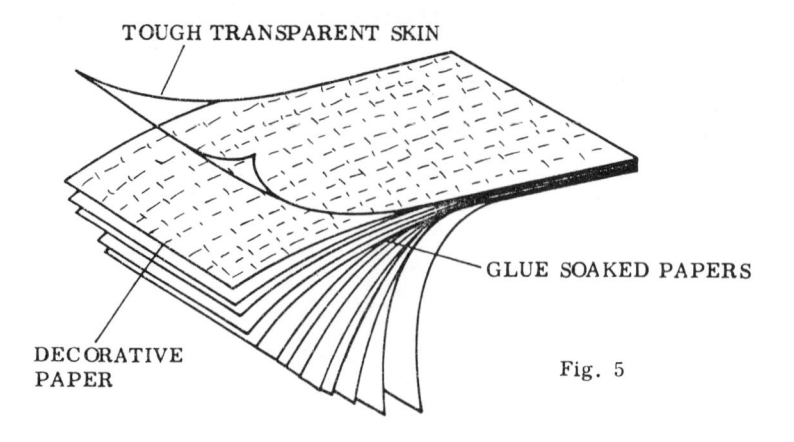

TOUGH TRANSPARENT SKIN

GLUE SOAKED PAPERS

DECORATIVE PAPER

Fig. 5

NPVC 204 PRO. SHEET MATERIALS.

BLOCK BOARD

This is a variation of plywood. A core of wood strips is glued together and faced with one or more veneers on each side (Fig. 1).

HARD BOARD

Low quality wood and wood wastes are ground and combined with water and glue to form a pulp. This mixture is spread between smooth aluminium sheets under great heat, forming a board. Hard boards have a smooth surface and a coarse side. They are available in thicknesses from 3 to 6 mm (Fig. 2).

CHIP BOARD

This is made from wood chips bonded together with glue. The chips are sorted, dried and mixed with the glue. Then they are spread on a plate and bonded with great heat and pressure. Chip boards are made in thicknesses from 6 to 60 mm (Fig. 3).

The edges of chip boards should always be protected, as they tend to split. This is done by glueing wood strips around the edges.

DECORATIVE LAMINATED PLASTIC

Laminated means consisting of a number of thin layers. Laminated plastics such as Formica (Fig. 4) are made by assembling many paper sheets soaked in glue. A decorative paper, also soaked in glue, is laid on top and over this is laid a transparent paper soaked in a very hard transparent glue which gives a tough surface. The assembled layers are placed between polished steel sheets and pressed at a high temperature (Fig. 5).

Decorative laminated plastics are durable, clean looking, smooth and attractive. They are made in a variety of patterns.

NOTES:

Fig. 1   Fig. 2   Fig. 3   Fig. 4

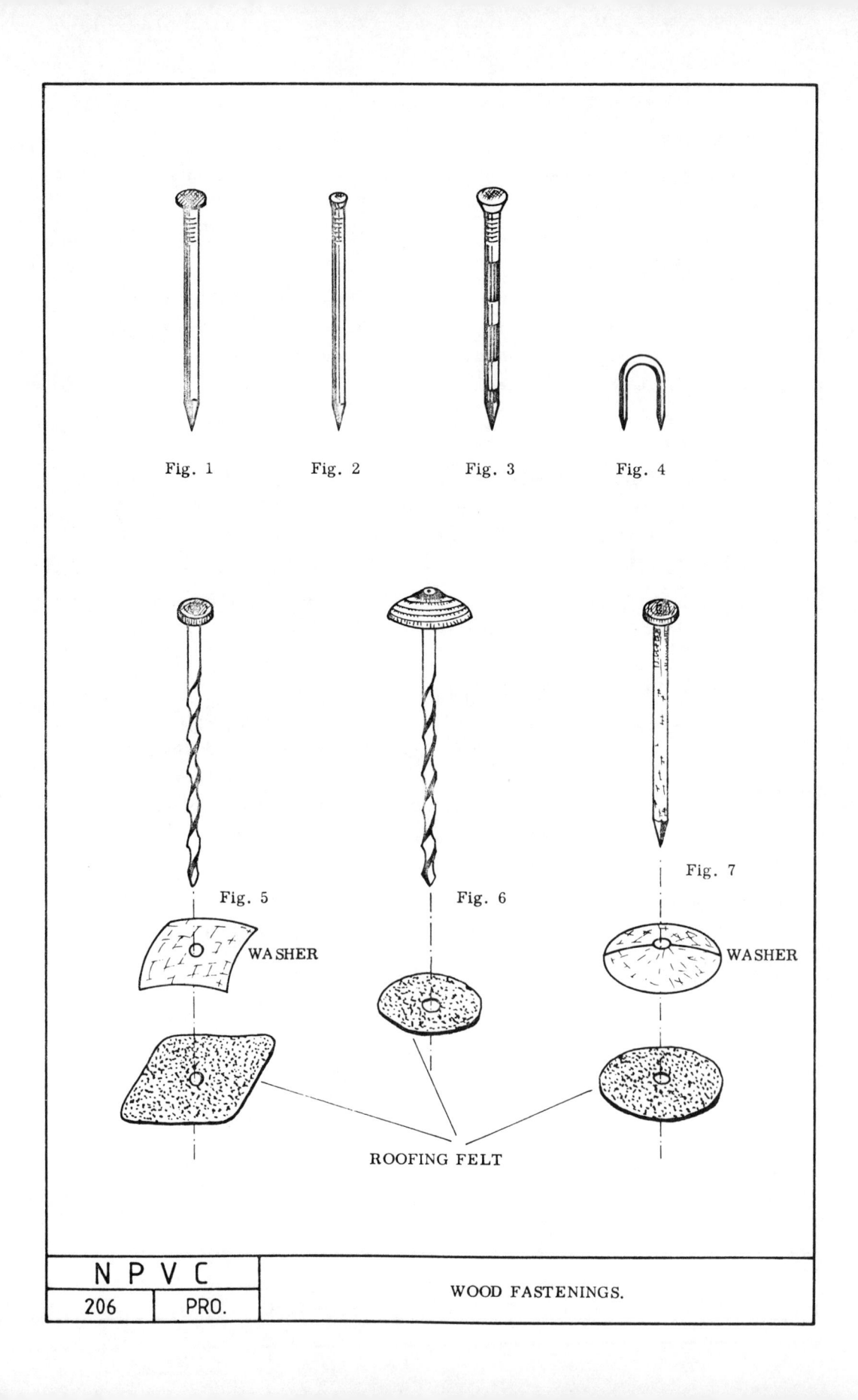

Fig. 5   Fig. 6   Fig. 7

WASHER   WASHER

ROOFING FELT

N P V C	
206	PRO.

WOOD FASTENINGS.

WOOD FASTENINGS

NAILS

Nailing is a fairly strong, cheap and quick method of fastening wood (see Rural Building, Basic Knowledge, page 92).

Nails have a head, shank and point and are usually made from mild steel wire. Galvanized, copper-plated or aluminium nails are used for work which will be in contact with water.

In Rural Building we use mostly wire nails. We group these into two classes: wire nails with flat heads and wire nails with very small heads, known as lost head nails. Some nails are used for special purposes, like staples, concrete nails and roofing nails.

- WIRE NAILS WITH FLAT HEADS: These are nails with large flat heads (Fig. 1). The head prevents the fastened member from being pulled off over the head of the nail.

  These nails are available in sizes from 7 to 310 mm long.

- WIRE NAILS WITH SMALL HEADS (LOST HEAD NAILS): These are wire nails with very small heads which can be punched or set below the surface of the wood and covered with putty. The disadvantage of this nail is that the nailhead can be easily pulled through the wood, so it cannot be used for heavy construction work (Fig. 2).

  These nails are available in sizes from 7 to 100 mm long. Small lost head nails are called panel pins.

- CONCRETE NAILS: These are hardened steel nails, available in different shapes and sizes. They are used to fix things to concrete or masonry (Fig. 3).

- STAPLES: These are U-shaped nails (Fig. 4) with two points. They are used to fasten wires and screens to walls or timber.

- ROOFING NAILS AND WASHERS: Special roofing nails are used for fixing corrugated sheet materials. They should be aluminium or galvanized metal to prevent rust, which could cause the nailhead to break off. The nails must be long enough to go at least 2 cm into the wood.

Drive screws, or screw nails as they are sometimes called, are commonly used for roofing (Figs. 5 & 6) and they have largely replaced nails with plain shanks (Fig. 7). The drive screws are galvanized and the shank is 2 mm or more in diameter, with a steep thread around it.

Drive screws can be driven in with a hammer, but they are very difficult to pull out again.

There are different types: either with a metal washer already attached to the head of the nail (spring head roofing nail, Fig. 6, page 206), or with loose metal washers (Figs. 5 & 7, page 206)

The washers or the spring heads prevent the nails from being pulled through the roofing sheet. They should be thick and wide enough (at least 2 cm in diameter) so that they secure the sheets well.

Roofing felt is always used under the washer to prevent leaking. The felt should be larger in diameter than the washer. Place the washers correctly: the hollow side should face the roofing felt and the sheet.

ORDERING NAILS

When ordering nails, state the kind of nail; the thickness (in 1/10 mm) and the length (in mm). Also state the material of the nails.

Example: Lost head nails; 16 x 30; steel

NOTES:

SCREWS

After nails, screws are the next most common type of wood fastener used in Rural Building. Screws are superior to nails because:

- they have greater holding power,
- they cause less shock to the work when driven into it,
- and they are easily removed, without damage to the work.

Screws are made of mild steel, brass, copper or they are galvanized. Usually mild steel screws are used because they are stronger than the copper or brass ones.

The parts of a screw are: the head (a), the slot (b), the shank (c), the thread (d), the point (e) and the core (f) (Fig. 1, page 210).

Screws are classified according to the shape of their heads:

- Countersunk head screws (Fig. 1, page 210)

- Round head screws (Fig. 2, page 210)

- Raised countersunk head screws (Fig. 3, page 210)

- Coach screws (Fig. 4, page 210).

Posidriv or Phillips screws have a head which is not slotted across the full width like common wood screws. They have a cross-shaped recess into which a special screwdriver fits (see page 76).

- COUNTERSUNK HEAD SCREWS: The head of this kind of screw is flat on top and tapering underneath. The length is measured from the point to the top of the head (Fig. 1, page 210).

  These are general purpose screws, used where the head of the screw must be flush with or below the surface of the wood.

- ROUND HEAD SCREWS: The head of these screws is round on top and flat underneath, and the length is measured from the underside of the head to the point (Fig. 2, page 210).

  These are used only where the head can be visible and can project above the surface of the wood, and when fixing light metal, where the metal is too thin for countersinking.

WOOD FASTENINGS.

a
b
c
d
f
e
length

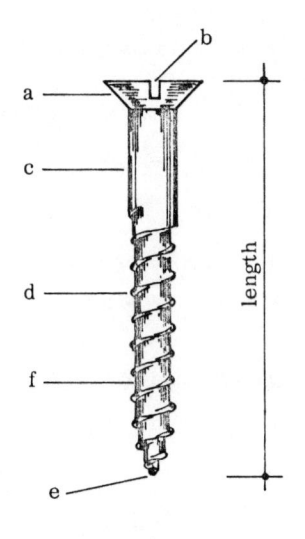

Fig. 1

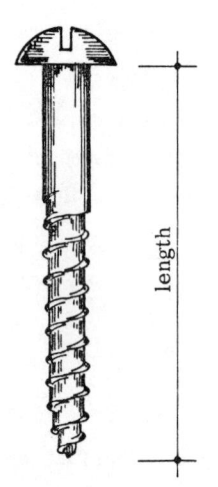

length

Fig. 2

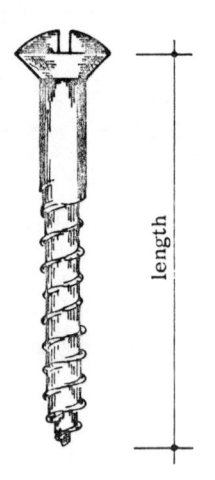

length

Fig. 3

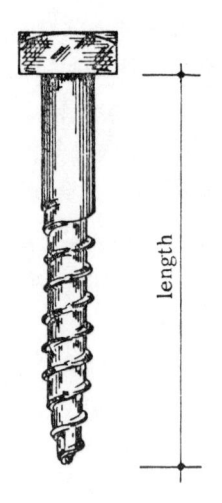

length

Fig. 4

N P V C		
210	PRO.	WOOD FASTENINGS.

- RAISED COUNTERSUNK HEAD SCREWS: The head of this screw combines the round and countersunk heads; round on top, tapering underneath. It is stronger than the round head screw because the head is less likely to break off. The length is measured as indicated in Fig. 3.

  These screws are used in fixing heavy fittings and thick sheet metal, where strength is needed.

- COACH SCREWS: This is a strong screw with a square head. Unlike other wood screws, it is turned with a spanner. Always put a metal washer under the head to prevent damage to the wood surface. The length is measured from the point to the underside of the head (Fig. 4).

  These are used for heavy construction work, for gate hinges, carriage work etc, where the head doesn't interfere.

ORDERING SCREWS

Screws are sold by number or in boxes containing a gross (144 screws). When ordering screws state the following in the order:

- thickness in mm
- length in mm
- kind of screw
- kind of metal
- amount needed

For example: 3 x 30; round head; brass screws; 3 gross   or,
5 x 50; coach screw, mild steel with washers; 2 gross.

NOTES:

WOOD FASTENINGS.

N P V C
PRO.   211

Fig. 1 — HEAD, WASHER, NUT

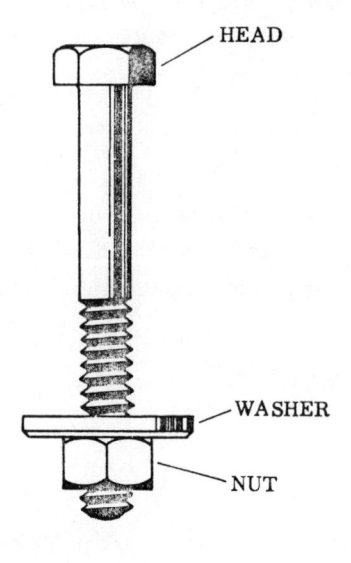

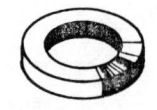

Fig. 3

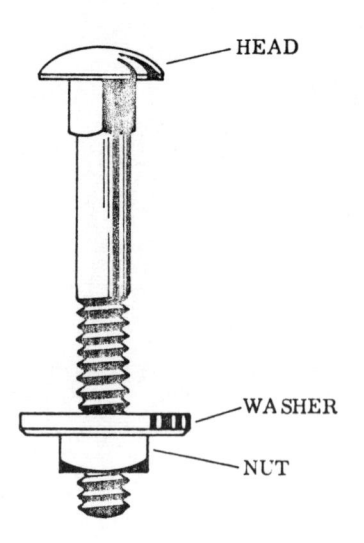

Fig. 2 — HEAD, WASHER, NUT

Fig. 4

NPVC		WOOD FASTENINGS.
212	PRO.	

BOLTS AND NUTS

Bolts and nuts are yet another means of fastening two pieces together. Bolts have hexagonal heads and are tightened up with the nuts. Bolts and nuts are used for heavy construction work (Fig. 1).

If bolts are used in timber, washers must be laid under the nut to prevent it from sinking into the wood (Fig. 3).

COACH BOLT

These bolts have oval heads and a square shank just under the head. This is so that the bolt grips the wood and doesn't turn when the nut is tightened up (Fig. 2).

The advantage of these bolts is that the head doesn't project up, since it is rounded and pulled into the wood.

WASHERS

A washer must always be used under the nut with both coach and regular type bolts. Never put a washer under the head of a coach bolt.

Washers can be made locally from a square piece of metal with a hole drilled in it.

SPRING WASHER

To prevent the nut from loosening when it is fastening metal to metal, a spring washer can be put between the metal and the nut. (Fig. 4).

NOTES:

CAN BE THREADED

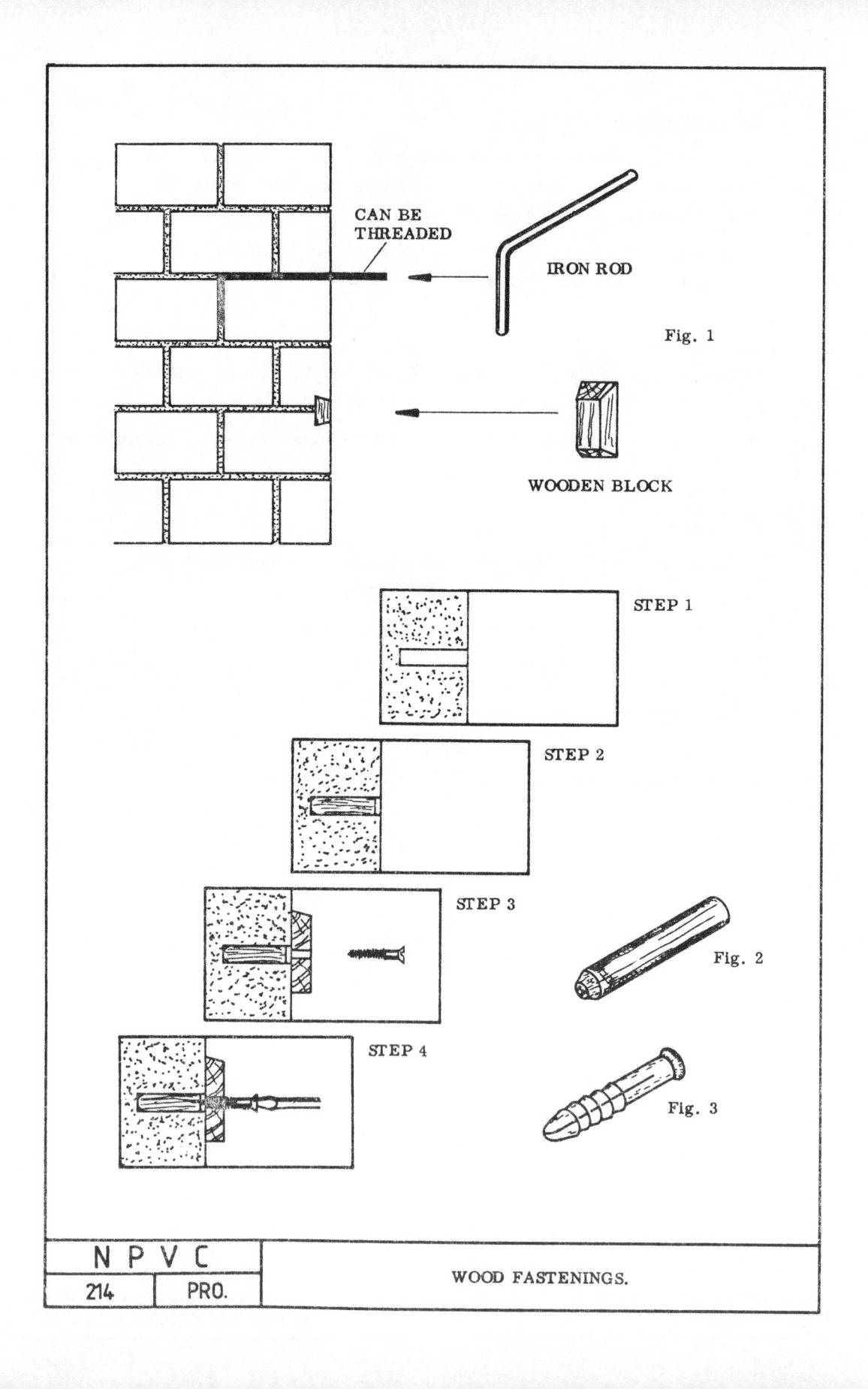

IRON ROD

Fig. 1

WOODEN BLOCK

STEP 1

STEP 2

STEP 3

Fig. 2

STEP 4

Fig. 3

N P V C	
214	PRO.

WOOD FASTENINGS.

ANCHORS

Fastening wood or other materials to concrete or masonry can be a problem. Often a screw will not hold in a landcrete block, or will not grip in concrete. Anchors can be used to solve this problem.

Anchors can be classed in two groups:
- anchors fixed during the initial construction and
- anchors which are fixed after the initial construction.

The first type of anchor can be an iron rod set into the wall during walling or casting, as is often done with the door frames. Threaded iron rods can be used, to receive a nut later on (Fig. 1).

Another way is to insert wooden blocks in the masonry; into which screws, etc. can be driven later. For maximum strength the wooden block should be dovetail shaped (Fig. 1) and it should be cut and fixed in a way that its shrinkage will have as little effect as possible on the wall.

The second group are the devices used to fix a piece to an already existing masonry or concrete work. The most simple of these is a wooden plug. A hole is chiselled or drilled into the masonry and into that hole is inserted a cylindrical plug of wood, which has the same diameter as the hole. The length of the plug should be a little less than the depth of the hole. The plug is made out of hard dry wood and the end which enters the wall is chamfered to enter smoothly. When a screw is driven into the plug, the wood will expand or even crack and the screw is wedged into position (Fig. 2).

More complicated devices, all sharing the same principle of holding a screw or nail by expanding, are now available. The most common one is a plastic plug (Fig. 3).

NOTES:

WOOD FASTENINGS.

Fig. 1

Fig. 2

Fig. 3

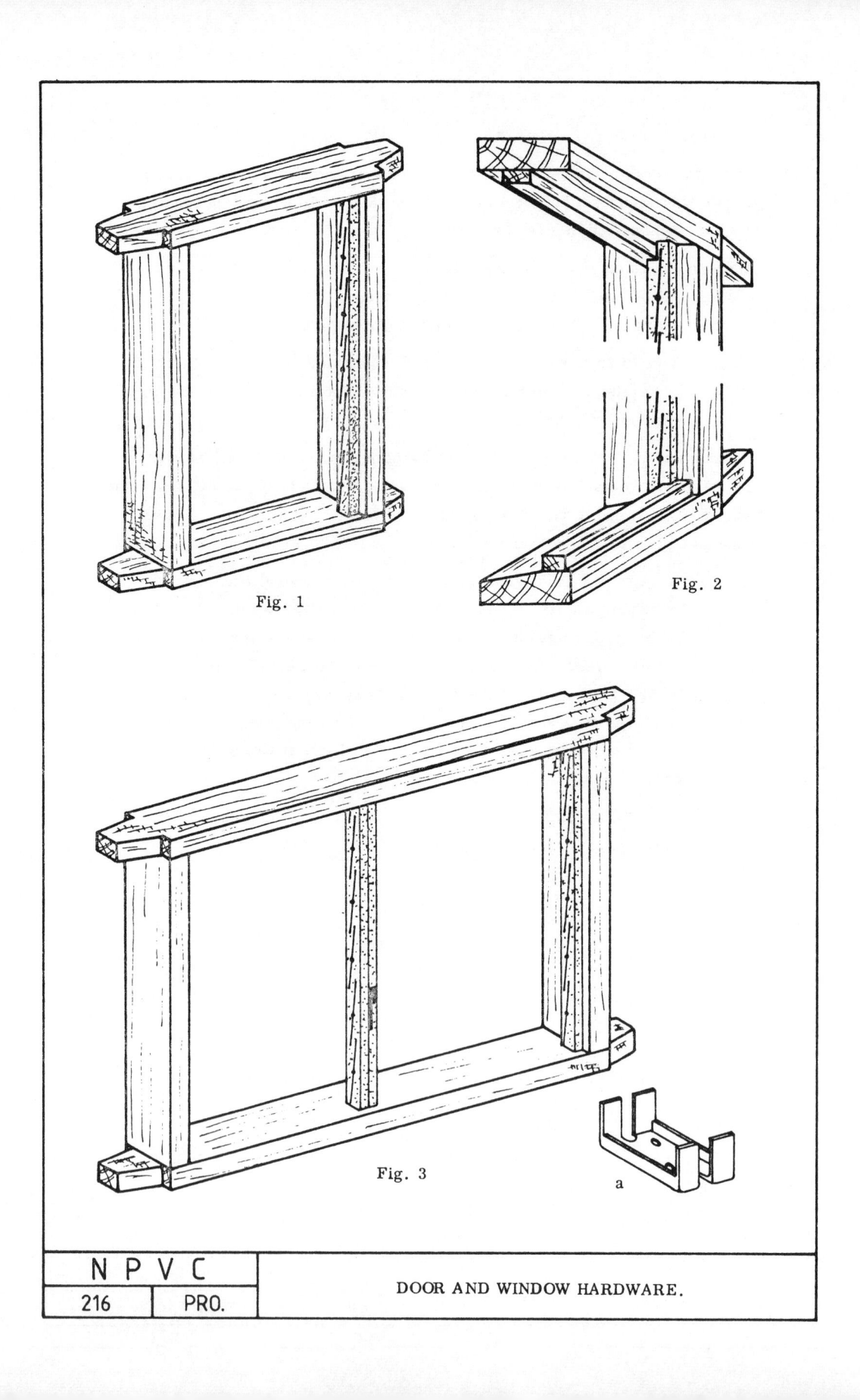

a

N	P	V	C	
216	PRO.			DOOR AND WINDOW HARDWARE.

DOOR AND WINDOW HARDWARE

LOUVRE WINDOWS

Windows with glass louvres are often used in the tropics because of their various advantages: they can be opened without any waste of space in the room and in the closed position they still admit light to the room. Their disadvantage is that it is very difficult to make them water-tight.

Louvre windows are installed ready-made into the window frame (Fig. 1). When the glasses are inserted in the window, wooden beads or ready-made aluminium waterbars are fitted to the head and cill to keep out dust and rain (Fig. 2).

In wide windows where two sets of louvres are installed in the frame, the metal posts of the louvre frame are fixed together in the middle, forming a metal mullion (Fig. 3). A separate wooden mullion is therefore not necessary. We will learn more in the Construction book (windows) about the installation of a self-mullioning louvre frame. Specially made mullion connectors are sold in a separate set (Fig. 3a).

Louvre windows are available in mild steel or aluminium frames. The aluminium frames need very little maintenance, but they are more expensive and less strong.

The size of the window frame is determined by the size and number of the louvre glasses. The inside width is:

- the length of the glass plus 3,8 cm (the thickness of two frames).

If two or more louvres are set across the width of the window, then for each additional glass 3,8 cm, plus the glass length must be added to the inside width of the frame.

The inside height of the window frame is determined by the number of louvre glasses.

Inside height (in cm) of window frames according to the number of louvre glasses:

No.	2	3	4	5	6	7	8	9	10	11	12	13	14
Ht.	30,5	44,5	58,4	72,4	86,4	100,3	114,3	128,3	142,2	156,2	170,2	184,2	198,1

All the above measurements are for 6 in. (15 cm) glasses. If you have other sizes, follow the manufacturer's instructions.

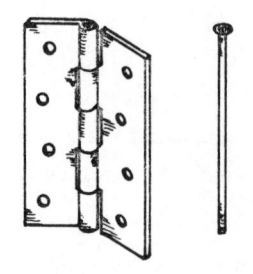

Fig. 1

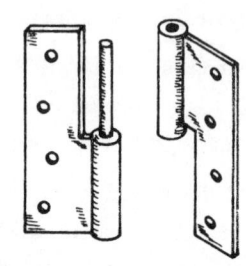

Fig. 2

Fig. 3

Fig. 4

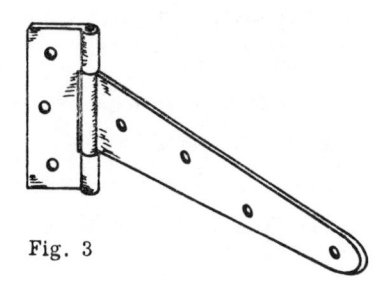

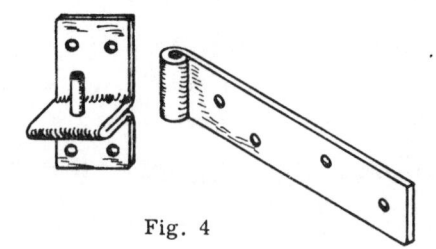

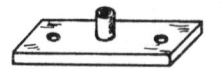

Fig. 5

N P V C		DOOR AND WINDOW HARDWARE.
218	PRO.	

HINGES

Hinges are available in almost countless different shapes, sizes and materials. The most common materials are steel, brass, and copper; or sometimes the hinge is only plated with brass or copper.

In Rural Building, we deal only with the most common types of hinges, which are butt hinges, H-hinges (Parliament hinges), T-hinges, band-and-hook hinges, and pivot hinges.

- BUTT HINGE: The ordinary steel butt hinge is cheap and durable, and it is the most common hinge for doors and casements (Fig. 1). It consists of two halves, also called leaves or flaps, held together by a pin. The pin may be removable or permanently fixed. If the pin can be removed from the outside when the door is shut and locked, the door is not burglar-proof.

When the door is shut, the two leaves (one attaches to the door post and the other to the hanging stile of the door) are folded together. The leaves are usually set into recesses in the door and post.

- H-HINGE (PARLIAMENT HINGE): The H-hinge, sometimes called the Parliament hinge, is similar to the butt hinge. It consists of two leaves, each with a knuckle. The pin is set permanently into the knuckle of one leaf. The H-hinge is installed in the same way as the butt hinge (Fig. 2).

- T-HINGE: T-hinges are mostly used for large and heavy doors, gates and ledged and battened doors. They are available in different sizes.

The hinge (Fig. 3) consists of a long mild steel strap, which is fixed to the outside of the door; and a cross bar which is hinged to the strap and attached with screws to the post of the door frame.

For security reasons, the strap of the T-hinge should be fixed to the door with at least one coach bolt, so that no one can unscrew the hinge to enter the building.

NOTES:

105 cm

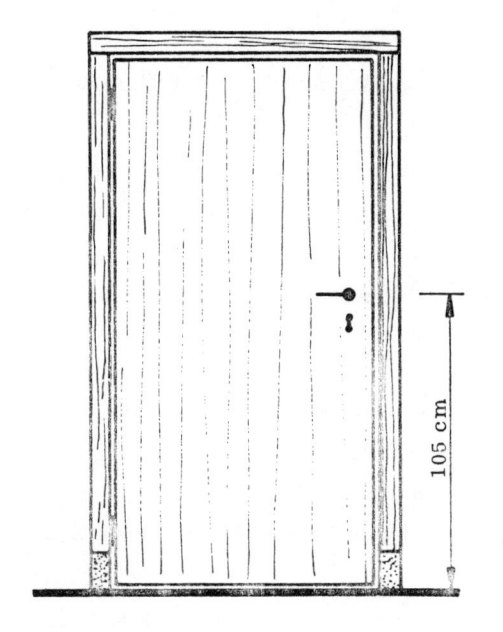

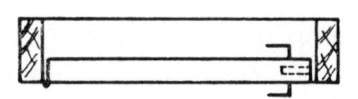

LEFT-HANDED

Fig. 1

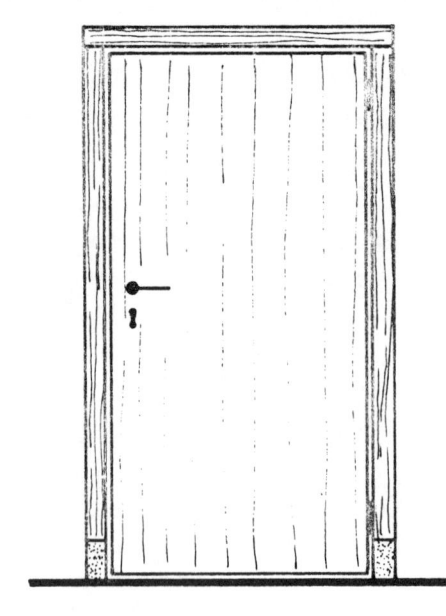

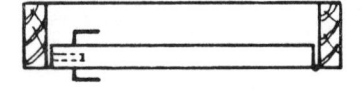

RIGHT-HANDED

Fig. 2

N P V C	DOOR AND WINDOW HARDWARE.
220	PRO.	

- BAND-AND-HOOK HINGE: The band-and-hook hinge is closely related to the T-hinge. It consists of an iron strip called the band, which drops onto a pin called the hook, which is attached to the frame of the door or window (Fig. 4, page 218).

They are used and installed in the same way as T-hinges.

- PIVOT HINGE: For windows, we sometimes use pivot hinges. These consist simply of a plate with a pin, which fits into a hole in another plate (Fig. 5, page 218).

LOCKS AND FITTINGS

There are many kinds of locks and fittings available for doors and casements. The choice between them depends on the type of door or casement and its function.

Doors and casements may be either left or right-handed. When the door opens towards you with the hinges on the left, it is a left-handed door (Fig. 1); if the hinges are on the right, the door is said to be right-handed (Fig. 2).

Some types of locks can be used on only one type of door, either right or left-handed. Therefore, we have to know whether the doors are right or left-handed before we order the locks, so we can buy the correct ones. Some types of locks have a latch bolt which can be changed to work in either type of door.

The most common types of locks are:

- the mortice lock
- the rimlock
- the padlock

The most common types of fittings are:

- the hasp and staple
- the barrel bolt
- the tower bolt
- the casement fastener

Locks are normally fixed at a height of 105 cm, measured from the floor to the centre of the handle (Fig. 1).

NOTES:

Fig. 1

Fig. 2

N P V C		DOOR AND WINDOW HARDWARE.
222	PRO.	

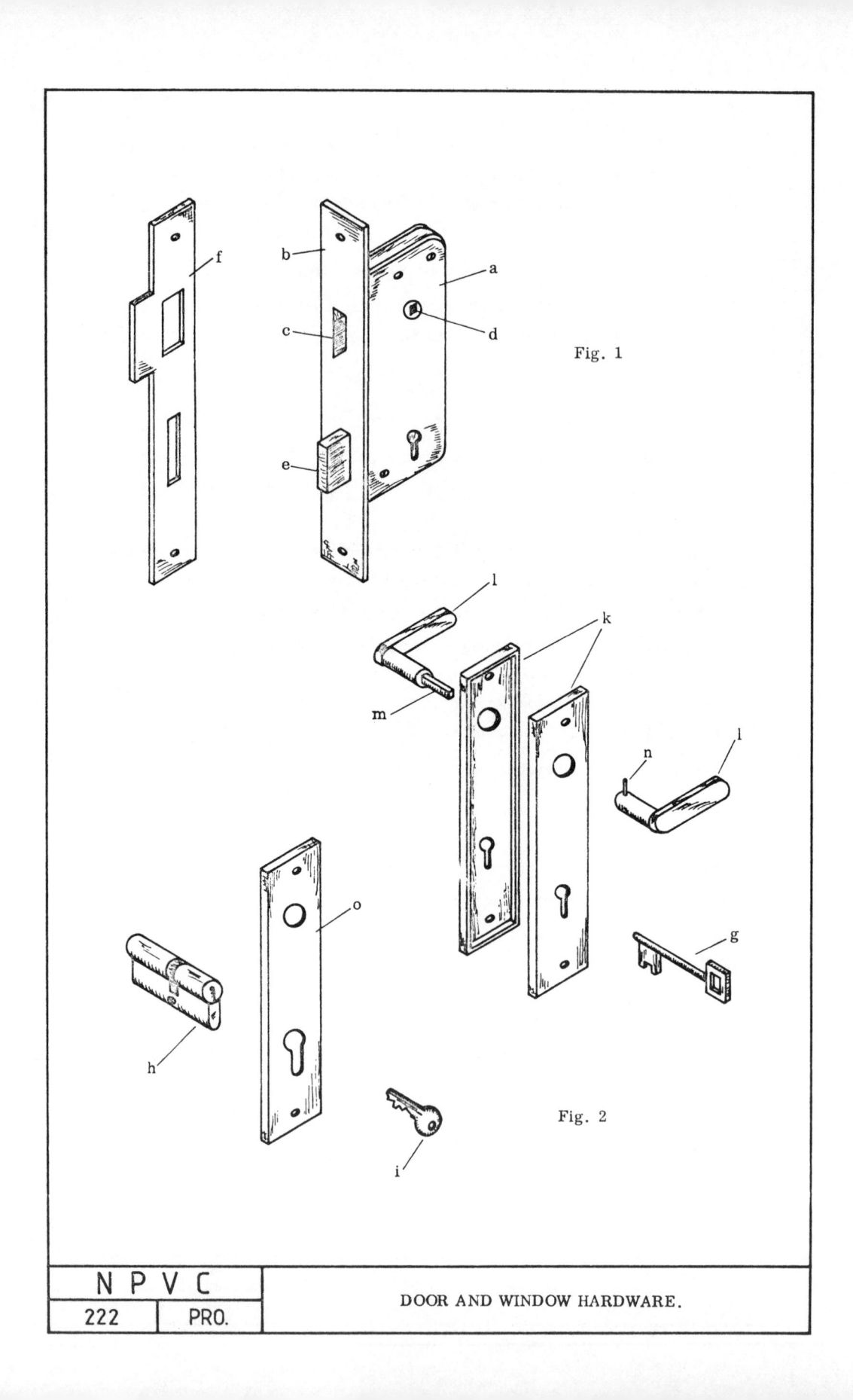

- MORTICE LOCKS: Mortice locks (Fig. 1) consist of a stock (a), faceplate (b) and the latch bolt (c) moved by the handle. The handle fits into the bush (d) and there is a lock bolt (e) moved by the key. The two bolts fit into holes in the striking plate (f) which is attached by screws to the door post.

As the name suggests, the mortice lock fits into a mortice in the edge of the door. The stock should fit tightly against the sides of the mortice so that the door itself takes the strain, not the screws which only hold the lock in position.

Mortice locks can only be installed in doors which are thick enough to receive a mortice. They are difficult to force open, since they are inside the door.

Mortice locks are locked either with an ordinary key (g), which moves small levers inside the stock to push the lock bolt in and out, or with a locking cylinder (h), which operates the locking mechanism inside the stock. The advantage of the locking cylinder is that it is more secure, since a special key is needed to open and lock it (i).

- DOOR FURNITURE FOR MORTICE LOCKS: The door furniture (Fig. 2) consists of two leaf plates (k), two handles (l) and a spindle (m). The spindle is permanently fixed in one handle and secured in the other by a pin (n). The handle with the pin should always be inside the door, so that the pin cannot be loosened from the outside.

The leaf plates are attached to both sides of the door with screws; or better, with specially made bolts which cannot be torn out easily. The leaf plates hold the handle in place and prevent damage to the keyhole. Sometimes separate leaf plates are used for the keyhole and the handle.

The leaf plate for a mortice lock with a cylinder (o) has an opening into which the locking cylinder fits, instead of a keyhole.

NOTES:

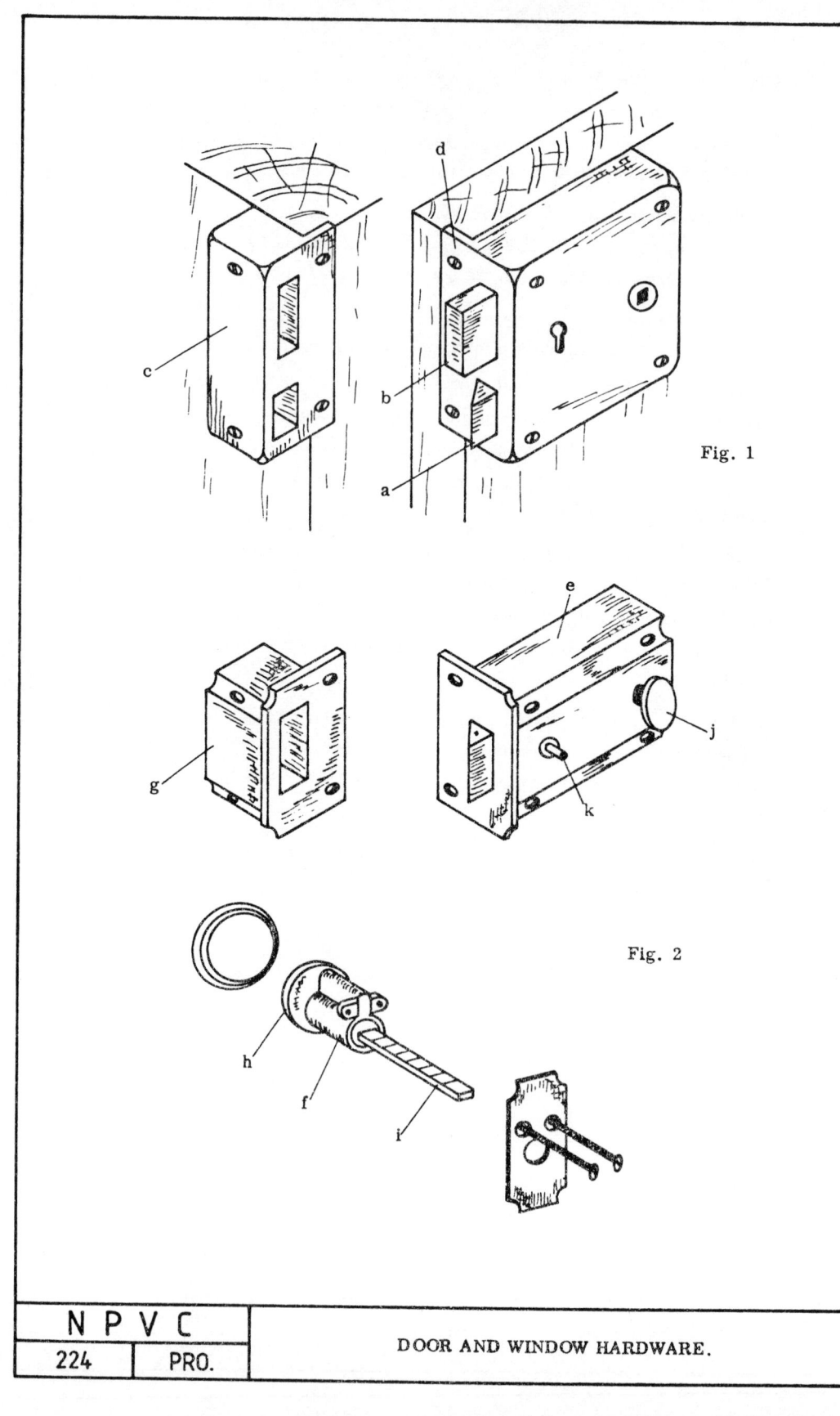

Fig. 1

Fig. 2

DOOR AND WINDOW HARDWARE.

NPVC 224 PRO.

- RIM LOCKS: Rim locks (Fig. 1) have a latch bolt (a) operated by a handle and a lock bolt (b) operated by a key from the outside or inside of the door. This kind of lock is attached with screws to the inside face of the door and the bolts shoot into a staple (c). There is a face plate (d) which is attached with screws to the door edge.

On the outside of the door, two round plates are attached with screws, one to hold the handle and the other to cover the keyhole. The handles have a square spindle which fits into the bush of the lock.

These locks are used on doors which are too thin to have mortice locks installed in them. Like the mortice lock, the rim lock is available with either an ordinary key locking system or with a locking cylinder.

- CYLINDER RIM NIGHT LATCH: This is a special kind of rim lock. It consists (Fig. 2) of a latch (e), a locking cylinder (f) and a staple (g). There is a face plate (h) which is part of the shell of the cylinder. The spindle (i) is fixed in the cylinder.

The latch bolt is operated from the outside by a key which rotates the spindle. The spindle moves the bolt mechanism inside the latch.

The bolt may also be shot back from the staple by turning the knob (j) of the latch from the inside. The locking arm (k) is used to fix the bolt in place, so that it cannot be operated from either side by the key or the knob, making the lock more secure.

- PADLOCKS: Padlocks have a ring which locks into a body. The locking mechanism can be either a lever mechanism or a locking cylinder, as with the mortice lock.

NOTES:

DOOR AND WINDOW HARDWARE.

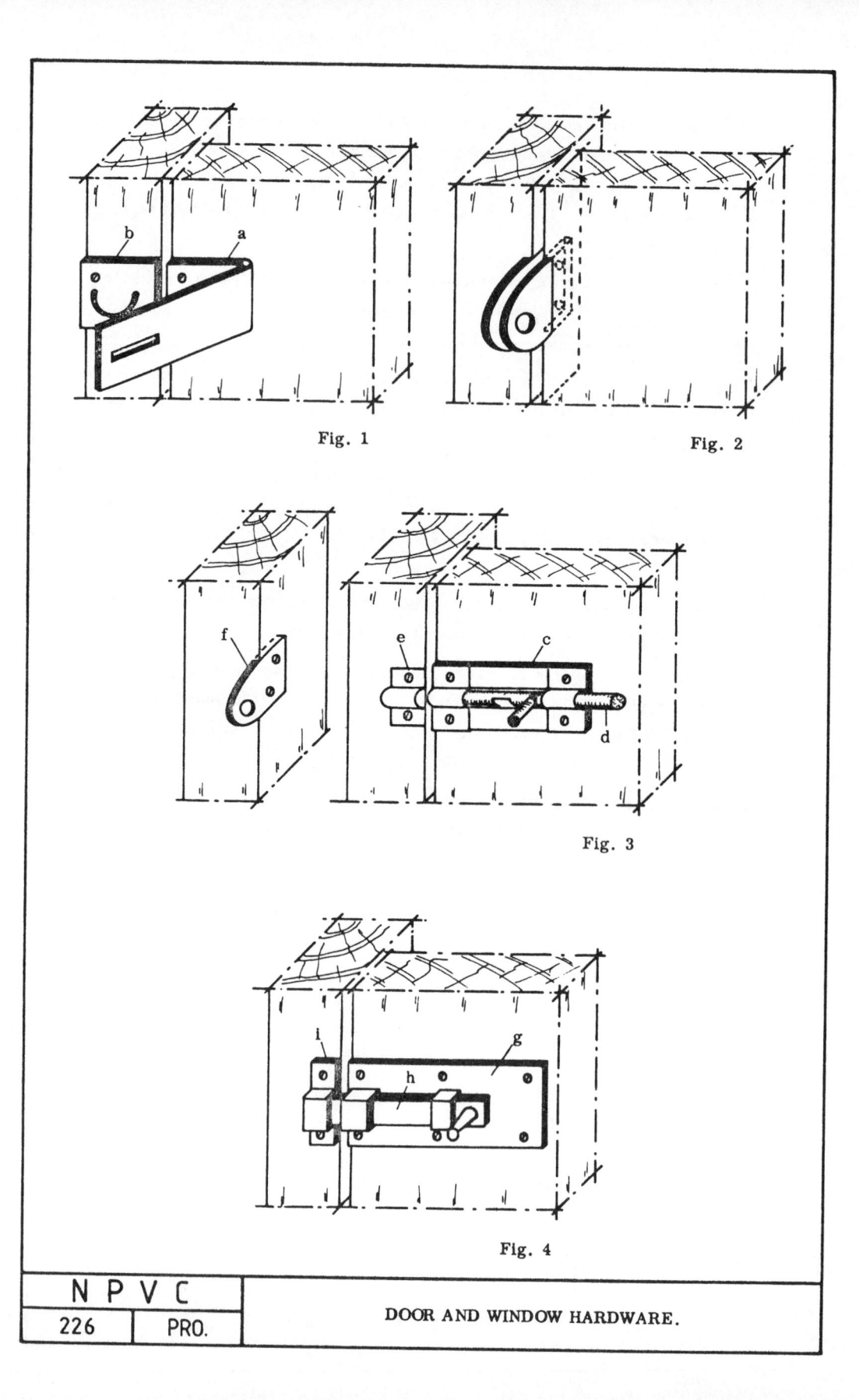

Fig. 1

Fig. 2

Fig. 3

Fig. 4

N P V C		DOOR AND WINDOW HARDWARE.
226	PRO.	

- HASP AND STAPLE: The hasp and staple (Fig. 1) is usually used in combination with a padlock. It is installed on doors which do not have to be opened very often, as it takes time to open it.

The hasp (a) is screwed to the door or casement and the staple (b) to the frame. Some types of hasp and staple can be unscrewed from the outside. For security reasons, these types should be attached with bolts and nuts rather than with screws. The leaf of the hasp should be installed so that when closed it covers the screws.

An alternative to the hasp and staple is the device shown in Fig. 2. This is an efficient way of locking doors with a padlock and the parts can be made by hand. Two small plates have holes drilled in them for the padlock and screws to pass through. The plates are attached with screws to the edge of the door and the post.

- BARREL BOLTS: The barrel bolt (Fig. 3) consists of a plate (c) with a round bolt (d). The bolt engages in a staple (e). The plate is attached with screws to the inside of the door or window and the staple is attached to the frame. Barrel bolts are often used to lock casements.

A more effective staple can be made by hand (f), in the same way as the device in Fig. 2 above. It provides more security than the staple normally supplied with the barrel bolt since the screws are on the inside face of the post and cannot be so easily forced out. The bolt plate can also be attached with bolts and nuts to make it more secure.

- TOWER BOLTS: A tower bolt (Fig. 4) consists of a plate (g) with a flat bolt (h) fitted in it. The bolt engages in a staple (i) or in a striking plate in the frame. the plate and bolt are usually fixed on the door or casement.

NOTES:

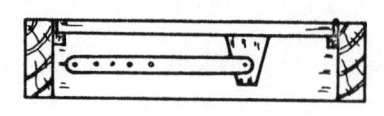

Fig. 1

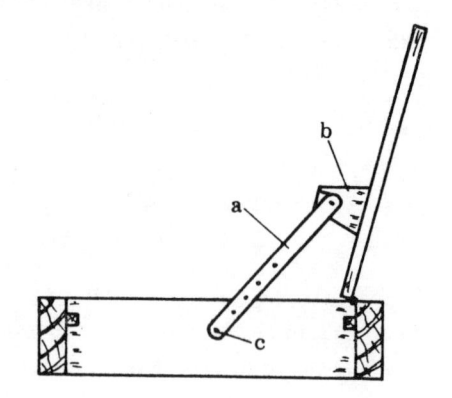

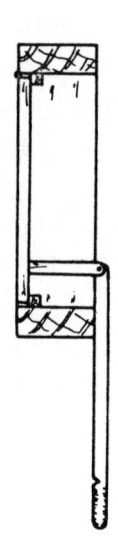

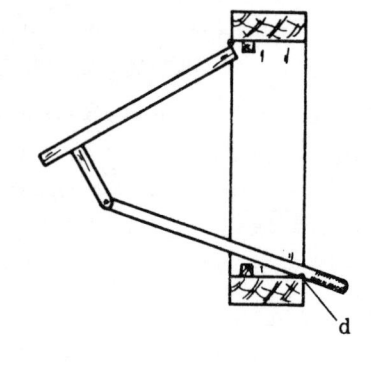

Fig. 2

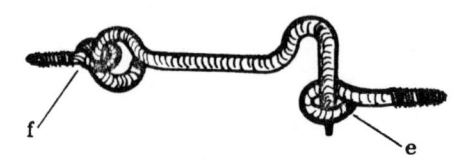

Fig. 3

NPVC
228   PRO.

DOOR AND WINDOW HARDWARE.

- **CASEMENT FASTENERS:** Besides the barrel bolt and the tower bolt, there are some other methods for keeping casements in a closed or open position. These are the casement stay and the cabin hook.

The casement stay serves to keep the casement in the open position. It consists of a handle (a), either wooden or metal, which is fixed onto a plate (b) screwed to the casement. A hole in the handle fits over a small pin in the cill of the frame when the casement is open and holds the casement in position (Fig. 1, c).

If the casement opens upwards, the stay can be constructed so that it also serves as a lock when the window is closed (Fig. 2). A notch (d) holds the stay on the cill when the window is open.

The cabin hook (Fig. 3) is used to hold the casement in the closed position. It consists of a hooked bar which fits into a screw eye (e). The other end of the bar is held by a second screw eye (f) which is fixed on the door or casement.

- **LOCKING DEVICE FOR LARGE DOORS:** The drawings below show a locking device used for large doors on stores etc, where trucks or other vehicles may have to enter to deliver goods (Fig. 4). It consists mainly of a long baulk (a) which can hold the two doors securely shut. The baulk is fixed with a coach bolt (b) onto one of the doors.

Two wooden blocks (c) hold the baulk in position when the doors are closed and a catch (d) prevents the baulk from swinging open by its own weight or from a gust of wind (Fig. 5). A piece of flat iron (e) bent as shown in Fig. 4, keeps the baulk in the locked position. This iron should be firmly attached to the door.

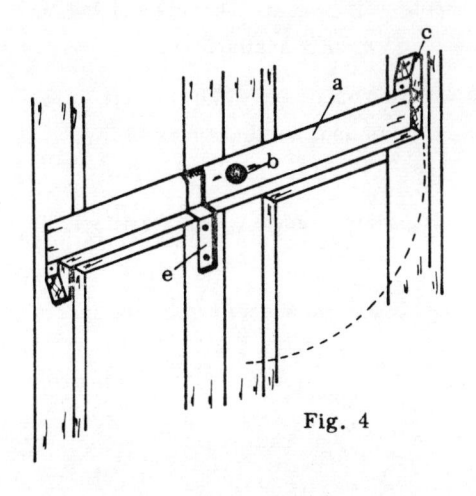

Fig. 4

Fig. 5

DOOR AND WINDOW HARDWARE.

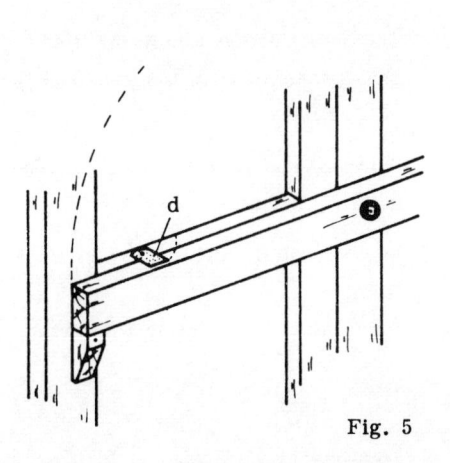

ROOF COVERINGS

When we talk about roof covering, we mean the non-load-bearing clothing of the roof. In Rural Building, we deal with three different types of roof coverings. These are:

- corrugated aluminium sheets,
- corrugated galvanized iron sheets and
- corrugated asbestos cement sheets.

All of these sheets are corrugated because that makes them stiffer so that they can go across the gap between the purlins without sagging.

CORRUGATED ALUMINIUM SHEETS

This roof covering is lighter and far more durable than the other types. Lightness is important so that the sheets can be easily transported. Aluminium sheets are rust-proof, easily installed and have a bright reflective surface which helps to keep the building cool.

However, the thinner gauge sheets are especially vulnerable to being dented, punctured, or torn off by storms. The sheets also creak a lot when temperature changes make them expand or contract.

Aluminium roof covering is available in a variety of shapes and sizes. The sheets can be from 61 to 76 cm wide and 200 to 400 cm long. Sheets up to 18 metres in length can be ordered from the factory.

In Rural Building, we mostly deal with sheets which are 61 cm wide and 244 cm long. This size is the most commonly used and readily available.

The sheets are available in different thicknesses or gauges. The thinnest is 26 SWG (Standard Wire Gauge; 26 SWG = 0,446 mm) and the thickest is 18 SWG (= 1,22 mm).

Store the sheets in a dry place. Avoid putting them in contact with fertilizer, lime or cement.

Further details and instructions concerning aluminium sheets can be obtained from:

    Ghana Aluminium Products Limited
    P.O. Box 124
    Tema

Also look at the Tables of Figures, pages 239 and 240, for more information.

CORRUGATED GALVANIZED IRON SHEETS

These are steel sheets which are corrugated and galvanized on both sides. They are usually thicker and heavier than aluminium sheets, so they lack some of the drawbacks of the aluminium, like being easily punctured or torn away in storms.

The most common size of these sheets is 61 by 244 cm.

CORRUGATED ASBESTOS CEMENT SHEETS

This kind of roofing sheet is also widely used. The disadvantage of these is that they are brittle and break easily if they are walked on or if a heavy object falls on them. They can also break if the roof construction warps due to the humidity changes in different seasons. Another problem is that they are heavy, and so they are difficult to transport.

The standard size is again 61 by 244 cm.

HOW TO ORDER SHEETS

When ordering sheets, you must keep in mind that the overlap of two corrugations on each sheet must be subtracted to find the effective width of the sheet.

RIDGE CAPS

Ready made caps for covering the ridge of the roof are also available. They are made from aluminium, galvanized iron or asbestos cement, to fit with the roofing sheet material.

Ridge caps can be made locally out of aluminium or galvanized iron sheets as will be explained when we come to roof construction.

The caps have to overlap the sheets on both sides of the roof. In the case of a roof which has the ridge along the direction of the prevailing wind (in northern Ghana, this is the east-west axis) the cap must overlap the sheets by 20 cm on each side. If the ridge lies across the direction of the prevailing wind (north - south) the cap has to overlap by 30 cm on each side (see Tables of Figures, page 239).

NOTES:

APPENDIX I : TABLES OF FIGURES

WEIGHT OF AGGREGATES

Dry sand weighs about 1800 kg per cubic metre, or 1,8 tons.

Broken stones weigh about 1700 kg per cubic metre, or 1,7 tons.

One slightly heaped headpanful of dry sand weighs about 30,5 kg.

One slightly heaped headpanful of broken stones weighs about 29 kg.

TABLE OF LOADING CAPACITIES FOR VEHICLES

Loading capacity	Maximum number of headpans	
	dry sand (fine and coarse)	broken stones (small and medium)
1 ton	33	34
1,5 tons	49	51
2 "	65	69
2,5 "	82	86
3 "	98	103
3,5 "	115	120
4 "	131	138
4,5 "	147	155
5 "	164	172
5,5 "	180	189
6 "	197	207
6,5 "	213	224
7 "	229	241
7,5 "	246	258

- REMEMBER: Never overload the vehicle! Observe the above quantities strictly.

NOTES:

TRANSPORTATION OF AGGREGATES

Supposing we have a 3,5 ton lorry available for transporting the aggregates, we can now figure out how many trips with the lorry will be necessary. If we know the total amount needed of each aggregate, we can find the number of lorry loads needed from the table on the left page.

For example, we need for the whole building: 429 headpans of fine sand;
152 headpans of coarse sand;
181 headpans of small broken stones;
165 headpans of medium broken stones.

The number of lorry loads required is found by dividing the number of headpans above by the number found in the table for the 3,5 ton lorry:

- Fine sand: 429 divided by 115 equals 3,73 or approximately 4 loads.

- Coarse sand: 152 divided by 115 equals 1,32 or approximately 1,5 loads.

- Small broken stones: 181 divided by 120 equals 1,5 loads.

- Medium broken stones: 165 divided by 120 equals 1,37 or approximately 1,5 loads.

- NOTE: It is always better to have a little extra material. Therefore, it is better to fill the lorry with every trip rather than take back a half-load. The cost of the transportation will be about the same anyway.

NOTES:

APPENDIX I : TABLES OF FIGURES.

BUILDING MATERIALS REQUIREMENT FOR ONE CUBIC METRE CONCRETE
(approximation values)

USE codes (key below)	MIX prop.	CEMENT bags (50 kgs)	head-pans	WATER buckets size no. 28	head-pans	AGGREGATES (in headpans) fine sand	coarse sand	small broken stones	medium broken stones
1	1:15	2	4,5	4,5	3	21	19	23	21
1;4	1:12	2,5	5,5	5,5	3,75	21	18,75	23	21
1;4;5	1:10	3	6,75	6,75	4,5	20,5	18,5	22,5	20,5
2;4;5	1:9	3,25	7,5	7,5	5	20,25	18,25	22,25	20,25
2;3;4;5	1:8	3,5	7,75	10	6,5	20	18	22	20
3;4;5	1:7	4	9	11	7,5	19,5	17,5	21,5	19,5
3;4;5	1:6	4,5	9,5	12,5	8,5	19	17,25	21	19
3;5	1:5,5	5	11,25	13,75	9	19	17	21	19
6;7;8	1:5	5,5	12,5	14,5	10	18,75	22,50	33,75	---
6;7;8	1:4,5	6	13,5	16,5	11	18,25	22	33	---
6;7;8;9	1:4	6,5	14,5	18	12	18	21,5	32,5	---
8;9	1:3,5	7	15,75	20	13,5	17,25	20,75	31,25	---
8;9	1:3	8	18	22	14,5	16,5	20	30	---

KEY FOR USE CODES

1 = foundations

2 = sandcrete blocks*

3 = mortar*

4 = plaster/render*

5 = floors

6 = columns

7 = beams/lintels

8 = slabs

9 = screed mortar*

- NOTE: Any reinforced concrete member of the structure must contain at least 270 kg of cement per cubic metre; and must be free of medium and large sized stones. Therefore, the figures in the upper part of the table are never used in mixing reinforced concrete.

* The items marked with a star are those which contain only sand as an aggregate. The amount of sand required is obtained by adding together the amounts of all the aggregates: fine and coarse sand, small and medium broken stones.

APPENDIX I: TABLES OF FIGURES.

HOW TO USE THE TABLE

When the plans for the building are completed, the builder can make calculations to find how much cement needs to be ordered. From the dimensions in the plan, the volume in cubic metres can be found for the various parts of the structure such as the foundation, floor, footings, etc. This is done by multiplying the width, length, and height of a part to get its cubic volume.

The examples below show how to calculate the cement required once you have found the volumes.

- FOUNDATIONS: Mix proportion = 1:10, volume = <u>5,75 cubic metres.</u>
- FOOTINGS: Mix proportion of sandcrete blocks and mortar = 1:8, volume = <u>3,3 cubic metres.</u>
- FLOOR: Mix proportion = 1:7, volume = <u>2,4 cbm.</u>
- CALCULATION:

　　FOUNDATIONS: According to the table, the mix proportion of 1:10 requires 3 bags of cement per 1 cbm. Therefore, we multiply 5,75 cbm x 3 = <u>17,25 bags of cement.</u>

　　FOOTINGS: The mix proportion of 1:8 requires 3,5 bags of cement per cbm. We multiply the volume of 3,3 cbm x 3,5 = <u>11,55 bags of cement.</u>

　　FLOOR: The mix proportion of 1:7 requires 4 bags of cement per cbm. We multiply the volume of 2,4 cbm x 4 = <u>9,6 bags of cement.</u>

We now add up the three results above and obtain a final result:　　17,25
　　　　　　　　　　　　　　　　　　　　　　　　　　　　　　　　　　+ 11,55
　　　　　　　　　　　　　　　　　　　　　　　　　　　　　　　　　　+　9,6
　　　　　　　　　　　　　　　　　　　　　　　　　　　　　　　　　　─────
　　　　　　　　　　　　　　　　　　　　　　　　　　　　　　　　　　38,40 bags

The total of 38,40 means that 39 bags of cement have to be ordered.

Do not forget to include all the members which contain cement in your calculations: landcrete blocks, mortar, lintels, concrete ring beam, etc.

The cement requirements for landcrete blocks varies according to the soil used. For approximate values see the table on page 179.

NOTES:

APPENDIX I : TABLES OF FIGURES.

When the cement requirements have been determined, we can use the table to find the quantities of aggregates that are needed. This is done by multiplying the same volume measurement by the appropriate number in the table.

- FOUNDATIONS:

    5,75 cbm x 20,5 equals approximately 118 headpans of fine sand.
    5,75 cbm x 18,5 equals approximately 106 headpans of coarse sand.
    5,75 cbm x 22,5 equals approximately 129 headpans of small broken stones.
    5,75 cbm x 20,5 equals approximately 118 headpans of medium broken stones.

- FOOTINGS: Since the sandcrete blocks and the mortar require only sand, all four quantities under aggregates (fine sand, coarse sand, small and medium broken stones) are added together and the result is multiplied by the volume of the footings: 20 + 18 + 22 + 20 = 80; 80 x 3,3 cbm = 264 headpans of sand.

- FLOOR:

    2,4 cbm x 19,5 equals approximately 47 headpans of fine sand.
    2,4 cbm x 17,5 equals 42 headpans of coarse sand.
    2,4 cbm x 21,5 equals approximately 52 headpans of small broken stones.
    2,4 cbm x 19,5 equals approximately 47 headpans of medium broken stones.

NOTES:

HEIGHT MEASUREMENTS FOR WALLS

This table can be used to calculate the height of a wall (not including copings, tie beams, or foundations) or to calculate the number of courses in the wall in order to make it a certain height.

LC = Landcrete wall; NC = number of courses; SC = sandcrete wall.

EDGEWISE LAID		
LC	NC	SC
24	1	25
48	2	50
72	3	75
96	4	100
120	5	125
144	6	150
168	7	175
192	8	200
216	9	225
240	10	250
264	11	275
288	12	300
312	13	325
336	14	350
360	15	375

FLATWISE LAID BLOCKS					
LC	NC	SC	LC	NC	SC
16	1	17	256	16	272
32	2	34	272	17	289
48	3	51	288	18	306
64	4	68	304	19	323
80	5	85	320	20	340
96	6	102	336	21	357
112	7	119	352	22	374
128	8	136	368	23	391
144	9	153	384	24	408
160	10	170	400	25	425
176	11	187	416	26	442
192	12	204	432	27	459
208	13	221	448	28	476
224	14	238	464	29	493
240	15	255	480	30	510

USING THE TABLE

- FOR EXAMPLE: The walls of a common house (not including wall plate or ring beam) are supposed to reach a height of 3 m above the finished floor level. We find with the aid of the above table that the plinth course of sandcrete blocks must be followed by 12 courses of landcrete blocks to reach a height of: 0,25 m + 2,88 m = 3,13 m. The floor is 10 cm thick, so the height above the finished floor level will be 3,13 m - 0,10 m = 3,03 m.

- ANOTHER EXAMPLE: The bottom of a water tank must be at a height of 4 m above ground level. With the aid of the table, the Rural Builder can easily figure out that he needs to make 24 courses of flatwise sandcrete blocks to reach this height.

APPENDIX I : TABLES OF FIGURES.

LENGTHS OF WALLS

LC = Landcrete blocks, length of wall (cm); No = number of blocks; SC = sandcrete blocks, length of wall (cm).

LC	No	SC	LC	No	SC	LC	No	SC	LC	No	SC
29	1	46	1114	36	1726	2199	71	3406	3284	106	5086
60	2	94	1145	37	1774	2230	72	3454	3315	107	5134
91	3	142	1176	38	1822	2261	73	3502	3346	108	5182
122	4	190	1207	39	1870	2292	74	3550	3377	109	5230
153	5	238	1238	40	1918	2323	75	3598	3408	110	5278
184	6	286	1269	41	1966	2354	76	3646	3439	111	5326
215	7	334	1300	42	2014	2385	77	3694	3470	112	5374
246	8	382	1331	43	2062	2416	78	3742	3501	113	5422
277	9	430	1362	44	2110	2447	79	3790	3532	114	5470
308	10	478	1393	45	2158	2478	80	3838	3563	115	5518
339	11	526	1424	46	2206	2509	81	3886	3594	116	5566
370	12	574	1455	47	2254	2540	82	3934	3625	117	5614
401	13	622	1486	48	2302	2571	83	3982	3656	118	5662
432	14	670	1517	49	2350	2602	84	4030	3687	119	5710
463	15	718	1548	50	2398	2633	85	4078	3718	120	5758
494	16	766	1579	51	2446	2664	86	4126	3749	121	5806
525	17	814	1610	52	2494	2695	87	4174	3780	122	5854
556	18	862	1641	53	2542	2726	88	4222	3811	123	5902
587	19	910	1672	54	2590	2757	89	4270	3842	124	5950
618	20	958	1703	55	2638	2788	90	4318	3873	125	5998
649	21	1006	1734	56	2686	2819	91	4366	3904	126	6046
680	22	1054	1765	57	2734	2850	92	4414	3935	127	6094
711	23	1102	1796	58	2782	2881	93	4462	3966	128	6142
742	24	1150	1827	59	2830	2912	94	4510	3997	129	6190
773	25	1198	1858	60	2878	2943	95	4558	4028	130	6238
804	26	1246	1889	61	2926	2974	96	4606	4059	131	6286
835	27	1294	1920	62	2974	3005	97	4654	4090	132	6334
866	28	1342	1951	63	3022	3036	98	4702	4121	133	6382
897	29	1390	1982	64	3070	3067	99	4750	4152	134	6430
928	30	1438	2013	65	3118	3098	100	4798	4183	135	6478
959	31	1486	2044	66	3166	3129	101	4846	4214	136	6526
990	32	1534	2075	67	3214	3160	102	4894	4245	137	6574
1021	33	1582	2106	68	3262	3191	103	4942	4276	138	6622
1052	34	1630	2137	69	3310	3222	104	4990	4307	139	6670
1083	35	1678	2168	70	3358	3253	105	5038	4338	140	6718

USING THE TABLE

Suppose you are planning a wall which is about 6 metres long, made from landcrete blocks. In this case it would be best to choose either 5,87 m or 6,18 m as the final length, so that it is not necessary to cut blocks to fit.

The table makes it simple to read off the number of blocks required per course, so it is easier to calculate the building materials that will be required. For example, a 34,70 m landcrete wall contains 112 blocks per course. The number of courses required to reach a certain height can be found in the table on page 237.

APPENDIX I : TABLES OF FIGURES.

ROOFING SHEET REQUIREMENT

This table is to help you to find the number of roofing sheets that need to be ordered, according to the length of the planned building. The number of sheets in this table refers only to the sheets along one side of the ridge line; so if you want to make a gable roof you have to double this number to get the number of sheets that will cover both sides of the roof. In addition, if the building is wide, so that more than one sheet is needed to cover the distance between the ridge and the lower edge; then the number from the chart has to be multiplied by the total number of sheets across the whole width of the building (see the example given on the next page).

The figures below are for the most commonly used roofing sheet size, which is 61 cm by 244 cm. The effective width of the sheets (minus 2 corrugations overlap) is approximately 50 cm (see pages 230 and 231). The figures include an allowance of 20 cm extra at each gable end.

NS = Number of roofing sheets; L (m) = the length of the building in metres.

NS	L (m)	NS	L (m)	NS	L (m)	NS	L (m)
3	1,1	14	6,6	25	12,1	36	17,6
4	1,6	15	7,1	26	12,6	37	18,1
5	2,1	16	7,6	27	13,1	38	18,6
6	2,6	17	8,1	28	13,6	39	19,1
7	3,1	18	8,6	29	14,1	40	19,6
8	3,6	19	9,1	30	14,6	41	20,1
9	4,1	20	9,6	31	15,1	42	20,6
10	4,6	21	10,1	32	15,6	43	21,1
11	5,1	22	10,6	33	16,1	44	21,6
12	5,6	23	11,1	34	16,6	45	22,1
13	6,1	24	11,6	35	17,1	46	22,6

SHEET REQUIREMENT FOR RIDGE CAPS

When the ridge of the roof is oriented in the direction of the prevailing winds (the direction from which the wind usually comes; in northern Ghana this is the east) then the ridge cap has to overlap 20 cm on each side of the ridge. Thus a common roofing sheet (244 cm long) will provide 6 ridge caps, each 40 cm long. To find the number of sheets required for the ridge caps, use the figure you get from the table above and divide it by 6.

For example: The building will be 10 metres long; looking in the table you find the figure of 21 sheets. Divide 21 by 6; this gives 3,5 (or approximately 4) sheets

APPENDIX I : TABLES OF FIGURES.

which are required for the ridge caps.

When the ridge of the roof is oriented across the direction of the prevailing winds (in northern Ghana this would be north-south), the ridge cap has to be wider so that the sheets are held better against the wind. In that case, the caps have to overlap each side of the ridge by 30 cm; therefore the roofing sheet can be cut into 4 pieces, each 60 cm long. To find the number of roofing sheets required for the caps, divide by 4 instead of by 6. Thus in our example, 21 divided by 4 gives 5,25 (or approximately 6) sheets required for the ridge caps.

ROOFING SHEET REQUIREMENT / WIDTH OF BUILDING

In order to avoid unnecessary waste, the Rural Builder should decide on the width of the building with the size of the roofing sheets in mind (see Construction, pages 158 & 159). The table below gives the span of the roof for a gable roof according to the number of sheets required to reach from ridge to eave level on each side of the roof. The sheets are standard 244 cm long roofing sheets, the roof has a 20 degree pitch, and there is an allowance of 50 cm extra on each side for the sheets to overhang (overhanging eave).

NS = Number of sheets; W (m) = Width (span) of the roof truss.

NS RIGHT	W (m)	NS LEFT
1	3,65	1
1	4,65	$1\frac{1}{2}$
$1\frac{1}{2}$	5,65	$1\frac{1}{2}$
$1\frac{1}{2}$	6,8	2
2	7,95	2
2	8,95	$2\frac{1}{2}$

NS RIGHT	W (m)	NS LEFT
$2\frac{1}{2}$	9,95	$2\frac{1}{2}$
$2\frac{1}{2}$	11,1	3
3	12,3	3
3	13,3	$3\frac{1}{2}$
$3\frac{1}{2}$	14,3	$3\frac{1}{2}$

- EXAMPLE: If the roof is to be 21 m long and 8 m wide (span of the truss), with a pitch of 20 degrees, and oriented with the prevailing wind, the number of sheets required can be figured as follows:

From the table on the preceeding page, we see that 43 sheets will fit along the ridge line. The table above shows that 4 sheets are required to cover the width of the house. When we multiply both numbers, we get the total amount of sheets which is 172 sheets.

We must also find the number of sheets required for the ridge caps. Dividing 43 by 6 gives approximately 8 sheets, making a total of 180 sheets to be ordered.

APPENDIX I : TABLES OF FIGURES.

APPENDIX II : GLOSSARY

Most new terms are explained in the text as they come up, but after that they are used without explanation. To make it easier for you, the words which tend to come up again and again are explained here once more, and references are given when possible to more thorough explanations in one of the text books.

- NOTE: The other books referred to here are given as abbreviations: the Basic Knowledge book is referred to as "BK"; and the Construction book is written as "Con". Where the page number only is given, it refers to the page in this book.

TERMS

- AGGREGATE: The sand, rocks or gravel which make up the greater part of concrete, sandcrete blocks, mortar, plaster or render (see page 147)

- ANCHORAGE: This refers usually to iron rods which are embedded in one part of the building and serve to hold another part in place, for example the roof anchorage which is embedded in the walls and holds the roof construction in place.

- ANGLE OF PITCH: This means the slope of the face of a saw tooth (see page 109); or else it refers to the slope of the roof construction (roof pitch), depending on the context where it is used (see page 158, Con).

- BATTEN: This usually refers to a small piece of wood that is used to help fix another piece, or it refers to the lengthwise boards of a battened door (see page 93, Con).

- BEVEL: A sloping edge; the edge of a piece which is cut off so the angle is no longer 90 degrees (see page 72, BK).

- BOND: This refers to the arrangement of blocks in a wall, in Rural Building usually a half-block bond (see page 8, BK).

- BRACE: A piece added to a construction to make it stronger or more stable.

- BUILDING UNIT: This is usually used with reference to blocks to specify the measurement of a block in the wall, including the thickness of one cross joint and the bed joint (usually 2 cm) (see page 4, BK).

- CAST: To pour concrete into a mold so that it hardens to a particular form.

- CAST-IN-SITU: This means that the concrete member is cast in its permanent position (page 170, BK).

- CHAMFER: An edge which is cut to a 45 degree angle; this is a special kind of bevel (see page 72, BK).

- CLEARANCE: Free space; or space to allow movement between parts.

- CLENCH: To secure a nail by bending the point over where it comes through the piece of wood (see page 94, BK).

- CONCAVE: This describes a surface which is hollow or curved inwards.

- CONCRETE: This is a mixture of sand, stones, cement and water which hardens into a rock-like substance (see page 166).

- CONVEX: This means that a surface is curved outwards, like the outer surface of a ball.

- COUNTERSINK: This means to enlarge the top of a screw hole so that the head of the screw will be flush with the surface (see page 96, BK).

- COURSE: This refers to a horizontal layer or row of blocks in a wall, including the mortar bed (see page 2, BK).

- CROSS-GRAINED: The wood fibres do not run parallel to the length of the piece of wood, so it is difficult to work with the wood.

- CURING: This refers to the process of hardening for any product which contains cement. The piece must "cure" for a period of days or weeks before it is ready to carry out its function.

- CUTTING ANGLE: The angle to which a cutting edge is shaped (see page 91).

- DIMENSIONS: The measurements of the length, width, and height of an object.

- END GRAIN: The end surface which is exposed when wood is cut across the grain (see page 72, BK).

- FACE SIDE OR FACE EDGE: This is the first side or edge to be prepared when wood is planed for use in a work piece, usually the best side or edge (see page 84, BK).

- FLUSH: When we say that two surfaces are flush, we mean that they are in the same plane, they form one flat surface together.

- FOOTINGS: These are the first courses of flatwise blocks which are laid on top of the foundations (see page 36, BK; also page 43, Con).

- FORMWORK: This is the wooden structure which holds and supports the concrete pieces while they are being cast (see page 170, BK).

- FOUNDATIONS: The solid base, usually concrete, on which the building rests. It is the only part of the building which is in direct contact with the ground (see page 15, Con).

- GAUGE: A measure, a means of comparing sizes (for example, see pages 13 and 39).

- GRAIN: The natural arrangement of the wood fibres (see page 72, BK).

- GRIND: To polish or sharpen by rubbing on a rough hard surface (see page 95).

- HEADER: This is a block which is placed in a wall in such a way that the smallest face is exposed (see page 2, BK).

- HONE: To give a final polished keen edge to a tool by rubbing it on the smooth side of a sharpening stone (see page 99).

- KERF: The cut made by a saw blade.

- LANDCRETE BLOCKS: This is a mixture of laterite soil, cement and water which is pressed into blocks in a landcrete block machine (see page 2, BK).

- LEVEL: A line or surface which is parallel to the horizon; horizontal. This can also refer to the tool, the spirit level (page 5) which is used to determine whether a surface is level.

- LINTEL: The wooden or reinforced concrete member which bridges the opening of a door or window at the top (page 164, BK).

- MARKING OUT: This means to make marks on a surface to show where later operations have to be carried out; for example marking out joints on frames or marking out foundations on the ground.

- MILD STEEL: This is the same as iron. Hardened steel is iron which has been hardened by a special process.

- MITRE: This refers to a joint where the two pieces are cut at a 45 degree angle so that they form a corner where the connection between them bisects the angle of the corner (see page 98, BK).

- MORTAR: A mixture of sand, cement and water which is used to form the joints between blocks; or as plaster or render, depending on the proportions of the mix (see page 158).

- MORTICE: A hole which is cut in a piece to recieve the end of another piece (see page 104, BK).

- PLASTER: The mortar layer which is applied to the inside walls of a building to make them smoother and more durable (page 174, BK).

- PLINTH COURSE: This is usually one course of edgewise laid sandcrete blocks which is laid on top of the footings (see page 36, BK).

- PLOT: An area limited by certain boundaries, it may contain one or more building sites (see page 140, BK).

- PLUMB: Vertically straight, perpendicular to the horizon.

- PRECAST: This means that the piece is cast beforehand and set into into its permanent position after it is hardened (see page 170, BK).

- PREFABRICATED: This means that the piece or part of the building is made ready before it is installed in its final position in the structure. In Rural Building we usually mean those things which are manufactured on the site such as doors, casements or frames. We also call this "ready-made".

- QUOIN: The outside corner of a wall (see page 26, BK).

- REBATE: A step-shaped rectangular cut in the edge of a piece (see page 72, BK).

- REINFORCE: To make something stronger by adding another material, for example the iron rods which are sometimes used to make concrete pieces stronger (page 171)

- RENDER: The mortar layer which is applied to the outside of the building to make the walls water resistant so that the blocks are not destroyed by rain (see page 174, BK).

- RIGHT ANGLE: A 90 degree angle.

- RIPPING: This means cutting a board along the grain, using a saw (page 45).

- SANDCRETE BLOCKS: These are blocks made with sand, cement and water, shaped in a sandcrete block machine (see page 2, BK).

- SCALE: In a drawing, this means the relation between the size of the drawing and the actual size of the object which is drawn; for example a building may be drawn in a scale of 1:100 (cm), which means that 1 cm on the drawing represents 100 cm in the actual building.

- SCREED: This refers to a strip of mortar which is laid on the wall to act as a guide during plastering (see page 176, BK); we also talk about floor screeds, by which we mean the 2 cm mortar layer which is laid on top of the base layer during floor construction (see page 179, BK).

- SETTING OUT: Marking the dimensions of an excavation with pegs and lines, or marking the positions of the walls on the footings, etc., sometimes called lining out.

- SHUTTERING: This means the parts of the formwork which are in contact with the cement (page 170, BK).

- SITE: This is the piece of land on which a building is made; a plot of land can contain one or more building sites (see page 140, BK).

- SOFFIT: This refers to the under-surface of a piece; in a concrete form, the bottom board of the shuttering is called the soffit board.

- SPECIFICATIONS: This means a detailed description of something.

- SQUARE: By this we usually mean that a piece has all its sides at right angles to each other, or that all the angles are right angles; or it can mean a rectangular surface where all the sides have the same length.

- STRETCHER: This means a block which is laid so that one of the long faces is exposed (either the top face or a long side) (see page 2, BK).

- STRUTTING: These are the supports which hold the shuttering in place when the concrete piece is cast (see page 170, BK).

- TAPER: This means that something is thinner at one edge than at the other.

- TRUE: This is a description meaning that a surface is completely straight and flat.

- VENTILATION: The air movement in and out of a room or area.

- WARP: This is any change of shape in a piece of wood which is caused by uneven shrinking or expansion (see page 132).

- WEDGE: A piece, usually V-shaped, which is used to hold other pieces under pressure, usually to hold pieces together as for example the handle and head of a hammer are fixed together by wedges (see page 41).

APPENDIX II : GLOSSARY.

NPVC